LIKE A VIRGIN

HOW SCIENCE IS REDESIGNING THE RULES OF SEX

AARATHI PRASAD

ONEWORLD

A Oneworld Book

Published by Oneworld Publications 2012

A CIP record for this title is available from the British Library

ISBN 978-1-85168-911-8

Ebook ISBN 978-1-78074-067-6

Cover design by Dan Mogford

Printed and bound by TJ International,
Padstow, Cornwall, UK.

Oneworld Publications
185 Banbury Road
Oxford
OX2 7AR
England

CONTENTS

For my mother, C. D. Nalini, my father, Tarran Persad Rambarran,
and my daughter, Sita-Tara

LIKE A VIRGIN

PROLOGUE

CONCEIVING THE INCONCEIVABLE

For most of human history, women have been given little credit when it comes to childbearing.

Sounds strange, I know. Women clearly carry babies and give birth to them, but you could argue, as many have through recorded history, that the female of the species merely serves as a 'vessel' – a kind of living incubator. This sort of thinking was by no means the preserve of the uneducated; the very reason it persisted as long as it did in mainstream cultures and medical practice was because it was 'tested', endorsed by the great scholars and shapers of medicine, from Plato through to Leonardo da Vinci. Plato appears to have made the claim that only men are complete human beings. Aristotle believed that men had the ability to generate a full human being, and that women were reproductively defective. Some four hundred years later, the physician Galen asserted that the female is imperfect compared with the male. Even when da Vinci made the first accurate drawings of a foetus in utero – sketching in chalk a single womb rather than multiple chambers, which were believed to give rise to twins – he still compared the growing embryo to the seed of a plant.

Such prejudices were handed down, generation after generation. Take, for example, Thomas Bartholin, a pioneering Danish scientist from an esteemed family of anatomists and medical scholars. Bartholin lived in the seventeenth century and discovered the lymphatic system – a finding that would have required a keen eye and the proficiency to carry out a detailed investigation of human anatomy. Yet, when it came to women and pregnancy, he also documented accounts of the birth of 'monstrosities' – such as the woman who delivered a rat, or another whose child had the head of a cat, because a cat had frightened her when she was pregnant. The idea that what a woman saw and felt, or that specific shocks or scares during pregnancy, would lead to specific defects in her baby was widespread. Being frightened by a mouse, for example, might lead to the baby having a mouse-shaped birthmark – or worse. Today, we would laugh at ideas like this, or dismiss them as urban legend. Why would a great scientist, an empirical type, treat any of these things as conceivable?

But then, many outdated ideas about sex and reproduction still persist in many places around the world. These beliefs at times prevent a woman from claiming full biological ownership of her child, though she is still culpable for any reproductive shortcomings (such as giving birth to a baby with defects; experiencing recurrent pregnancy loss; failing to get pregnant). Even in our genetic age, there are women who are blamed (and who blame themselves) for giving birth to girls instead of boys. On the other hand, while the scientific evidence is still mounting, it appears to be true that a mother who, for instance, suffers stress during pregnancy will leave a lifelong mark on her child.

And while we all know that the sperm determines whether a child is a boy (XY) or a girl (XX), new research shows that a woman's immune system screens sperm after they have entered her body, and some women's bodies are more likely to discard Y-carrying, boy-making sperm. Many of these ideas, based in fact or fiction, are descendants of the cultural vocabulary of ancient Greece and the Renaissance. Despite the reality of test-tube babies and sperm banks, it seems we haven't moved much beyond Bartholin's theory of a rat-child or blame being placed at a mother's feet.

Far stranger, however, is another long-lived belief: the concept of a virgin birth. From the fertilization-by-feather of the Aztec Coatlicue to Isis's recipe for resurrecting her dead husband Osiris's phallus, to the Blessed Virgin Mary, there appears to be no culture that does not embrace some legend of a woman giving birth without mortal man. You could imagine that because, in the view of classical thinkers, women were viewed as crude vessels for reproduction, a repository for the vital semen of man, that they could just as easily have their bodies appropriated by their gods (who, of course, are generally male). But the myths extend to active agents that are not all divine. Human virgin births have been said to be caused by such things as sunlight and eating magical fish. These examples illustrate just how compelling was the notion of impregnation without sex, in part because there was so little real understanding, for so long, of how babies are actually made.

We now know that a mother makes an essential contribution of genetic information to her child, as well as providing a protected environment and the physical building blocks for the embryo's developing body. Biologically and genetically, women clearly are not mere vessels, nor are they redundant. Still, even when confronted by the double helix of DNA, the combination of the sex chromosomes X and Y, and genetic variants and mutations, the belief in the possibility of virgin birth has proved surprisingly enduring. It has ranged from the technology-fuelled optimism of the post-war boom, when doctors hunted for a virgin mother via the tabloids; to the absurd insurance policy, offered in the past decade, that would cover the cost of bringing up a child should you experience a virgin birth.

The simple truth, for humans at least, is that neither women nor men are currently redundant when it comes to making babies. Though the females of many animal species have the option of reproducing quite on their own, for us, a mutual need was established in our distant evolutionary past, and once that treaty was written in DNA, it could never be broken. Never, that is, until now.

In the future, technology might vindicate a few of the ancient, seemingly absurd concepts – at least in some respects. It might someday be possible to create a child from one parent alone. Ironically, because of the way men and women's chromosomes are arranged, the virgin parent will more likely be a man. Geneticists are cracking the codes that block our eggs from becoming embryos without sperm; stem cell scientists are creating eggs and sperm from bone marrow; artificial wombs are being built; artificial chromosomes are being constructed. And it seems not a minute too soon. Both male and female infertility is on the rise, and some scientists are warning that the Y chromosome, the very thing that makes men both fertile and male, is slowly but surely dying; it now has only around forty-five genes of the 1400-odd genes with which it began the human species. If the Y chromosome's genetic information essentially disintegrates, what solution could technology offer to sustain, well, *us*?

This is more than a question of futuristic science. The gender roles assigned to us by the fact of sex have, over the centuries, been used to oppress women and justify anti-homosexual prejudices. If we can make babies without sex, the family structures that we've come to view as traditional may well change beyond recognition. All of these attitudes will need to be reconfigured to fit our new lives 'after sex'.

In this way, the ancient myths of a virgin birth may prove to be the prehistory of our species' future. As Miss Miniver exclaims in H. G. Wells's feminist novel, *Ann Veronica*: 'Science some day may teach us a way to do without [men]. It is only the women matter. It is not every sort of creature needs – these males. Some have no males.' To which Ann Veronica replies, with some hesitation, 'There's green-fly.'

I began to wonder about these paradoxes, ironies, misnomers, interpretations, and reinterpretations of the reproductive role of women. Are the ideas of the ancients all myth, and all those of modern biology fact? What does the future hold in store? What will we face if we start making babies like a virgin? Will we ever be able to return to sex, and do we even have the choice?

PART I

THE MYTH OF THE NATURAL BIRTH

Sit down before fact as a little child,
be prepared to give up
every preconceived notion

Thomas Henry Huxley

1
PLANTING THE SEED

We must first establish 'how' in order to know whether or not we should be asking 'why' at all...
Stephen Jay Gould, *Natural History*, 1987

On 28 October 1533, the fourteen-year-old Catherine de Medici married the fourteen-year-old Henry, the Duke of Orléans. Catherine brought a substantial chunk of the Medici family fortune to France as her dowry, but as soon as it became clear that her husband would rise to become King Henry II, her true value was seen to be in her womb, in which she would produce the nation's heirs.

Over the following ten years, however, Catherine failed to become pregnant. This was not for want of trying. A dispatch to the Milanese government reported that her father-in-law, Francis I, had made a point of watching the royal couple in their bed to make sure the union was consummated – and was pleased to observe that each 'jousted valiantly'. As attempt after attempt failed, rumours of an imminent divorce spread through the court. Catherine promptly surrounded herself with doctors, diviners, and magicians. She refused to travel by mule, believing that the infertile beast would transmit its sterility to anyone who rode one. She consulted tarot cards, charms, and alchemy. She drank the urine of pregnant animals; ate the powdered testicles of boars, stags, and cats; dutifully swallowed cocktails of mare's milk, rabbit's blood, and sheep's urine. Catherine's sterility was torture to her.

But the young queen was not alone. Henry's lifelong mistress, Diane de Poitiers, never bore him a child, even though she was already a mother of two. Though she remained an exceptional beauty throughout her life, Diane was nineteen years Henry's senior, well past peak fertility at the time their love affair began. She knew Henry better than anyone, even Catherine, who was atrociously envious of the king's mistress. Diane's advice was that Henry and Catherine should make love *à levrette*, in the style of a greyhound bitch. She likely suspected that Catherine was perfectly capable of getting pregnant, and her advice was not unfounded: she knew that Henry's genitalia were misshapen, from a condition known to doctors as hypospadias, in which the urethra develops abnormally. But then, Diane was not the only person to know of Henry's affliction. As one seventeenth-century biographer put it:

> It is sufficient to say that the cause [of infertility] was solely in Henri II… nothing is commoner in surgical experience than such a malformation as the prince's, which gave rise to a jest of the ladies of the court.

The odd position of the opening of Henry's urethra appears to have twisted his penis into a downward curve. Chances are that he simply couldn't get the royal semen to where it needed to be. Yet, Catherine got the blame.

⦾⊙⦾

Throughout human history, our understanding of how babies are made has been draped in layers of myth and assumptions, many quite heavily stained by the politics of gender. Human dissection was taboo for most of recorded history, and effective microscopes would not be fabricated until the seventeenth century. For millennia, it was not easy to figure out what was really going on inside a pregnant woman's body, which made it much easier to assume that what was happening there was either miraculous or meaningless.

Fertility appears to have been among the earliest concerns of the earliest humans. In Bronze Age societies, these reproductive affairs were viewed simply: by some form of magic, a woman grew large, and out of her body came a child. It was women, not men, who were worshipped as the givers of life; women who were placed on a pedestal for their seemingly miraculous powers. Some of the very earliest objects of worship found by archaeologists working around the Mediterranean are wide-hipped, corpulent-bellied, ample-breasted figures: unmistakably female. In some cases, these figures appear to have once held opium-rich poppy heads – an invaluable panacea, often used to ease the excruciating pains of childbirth. One such statue, carved from mammoth ivory, has been dubbed the Venus of Hohle Fels; discovered in 2008, it is thirty-five thousand years old, the oldest known figurine representing any human form. These early artworks, and the Venus of Hohle Fels in particular, emphasize our external sexual organs, that is, how sex works, superficially.

The next great breakthroughs in exploring the mechanics of sex are found housed in the archives of Tehran University. There, the catalogue lists one of the few remaining manuscripts of the *Kitab al-Hayawan*, or *Book of Animals*, by the ninth-century Muslim scholar al-Jahiz – a document too delicate for any but the most circumspect of scholars to handle. In this great work describing hundreds of animal species, al-Jahiz included a volume, then only recently translated into Arabic, entitled *On the Generation of Animals* by the philosopher Aristotle. Aristotle's tract had been salvaged from near oblivion by the physician Thabit ibn Qurra, who wrote widely on medicine, astronomy, and mathematics. Al-Jahiz and Thabit were part of a group of medieval Islamic thinkers who, through the darkest ages of European science, preserved, utilized, and developed the medical ideas that had been elaborated centuries earlier by the Greek masters – Hippocrates, Plato, Aristotle, and Galen. Thus, when in the sixteenth and seventeenth centuries Western Europe emerged into the Renaissance, the biological concepts they resurrected belonged squarely in the third century BCE – including the belief that reproduction was predominantly a male affair.

This idea was not a completely new one, even in the third century. The Egyptians and the Indians as far back as the fourteenth century BCE described a man's contribution as the seed sown in the fertile ground of a woman's body. The great Greek dramatist Aeschylus, in his tragedy *The Eumenides* from 485 BCE, defines a parent as 'he who plants the seed. The mother is not the parent of that which is called her child but only nurtures the new planted seed that grows.' Even following that line of thought, men and women should have held the same reproductive value, because women were, in theory, still required. But a parent was the person who planted the seed, which meant a woman could only play the role of nurse.

Aristotle was the son of a doctor, so he may have been familiar with these common conceptions long before he attended Plato's Academy in Athens to study philosophy and science. Around the time of his teacher's death, in 347 BCE, Aristotle moved to Assos, in Turkey, to set up his own school, and then moved on to the neighbouring island of Lesbos, where he became tutor to the son of King Philip II of Macedon, later Alexander the Great. Inspired by Aristotle's teachings, Alexander was inclined towards medicine, but he eventually preferred conquering the world. Once his teaching assignment was fulfilled, the master returned to Athens and sat down to complete his book on the animals. In it, he covered a massive amount of ground, including the origin of sperm, the causes of pregnancy and infertility, and the purposes of menstruation and lactation.

From the outset it was clear to Aristotle that semen was the male contribution to making a baby. In trying to pinpoint the female equivalent, he landed on menstrual blood. In Aristotle's well-honed reasoning, both ejaculation and menstruation appeared during adolescence. He also observed, perhaps from home experiments, that after repeated ejaculation semen became bloody; thus, like a woman's monthly period, semen, too, must be made out of blood. As far as Aristotle was concerned, each animal could only have one kind of bodily fluid from which to make babies. Because the female had bleeding, she could not have semen – or something else that contributed to the creation of children.

However it was that Aristotle conducted his research, he was aware that a woman didn't just bleed; she could at times also release a clear fluid during sex. He resisted the idea that this fluid might contribute to reproduction in the way that semen did, since the part of the woman that experienced pleasure from sexual contact was not the part from which this fluid was released. In any case, if a woman had her own semen, then she really should be able to make babies without a man, a hypothesis for which he had no evidence – at least not in humans.

He suspected some female animals could have babies without

males, and noticed some animals had no males or females, that is, no sexes at all. But Aristotle may have worked out this theory by observing animals in which it is extremely difficult to tell the males and females apart, just by eye. For instance, some vultures, where the males and females have identically coloured feathers so that the sexes appear exactly the same, as opposed, say, to peacocks and peahens, where the sex is very evident.

While living on Lesbos, Aristotle had made sure to include hyenas in his animal studies. His great interest in the animal had been piqued by the rumour that 'every hyena is furnished with the organ both of the male and the female' – that they were hermaphrodites. Today, the reason for the rumour is plain: the female hyena has a clitoris so grossly enlarged that it looks, to the casual observer, much like a penis, especially when the clitoris is fully erect, when it can protrude to seven inches. Spotted hyenas are, in fact, the only female mammals that urinate, mate, and give birth through the tip of a clitoris. (Keep in mind that the hyenas give birth to infants that weigh between 1 and 1.5 kilograms – and sometimes to two infants at once.) The female hyena lacks an external vagina; in place of the labia majora, the fleshy folds that normally flank the vagina, it has a fused sac of skin, something like a scrotum. If you were to look inside the female's 'penis', however, you would find a urinary and genital system far more typical of any other female mammal.

Aristotle studied his hyenas carefully. His were not the spotted variety, but striped, as were found throughout the Mediterranean region of his day. And like their spotted cousins, male and female striped hyenas look remarkably similar. Both have manes that are erected when the animal is threatened – manes so large that Aristotle described them as running 'all along the spine'. The females had the same enlarged, penis-like clitoris as the spotted hyena, and the males appeared to have a large opening near the anus, looking much like a vagina. When it came time to dissect the specimens of hyena that he had collected, Aristotle soon realized that the rumour that the animals were hermaphrodites was untrue. In addition to noting the differences between the clitoris in the female and the penis in the male, he identified the opening in the male's anus as a sweat gland. By virtue of its position, this structure, he explained, could easily be confused with a vagina. Behind the opening, however, he did not observe any plumbing that might allow it to be used as a passage through which fertilization might happen.

It was an obvious case, in Aristotle's view, of mistaken identity, the external appearances hiding the significant differences in what was going on inside the animals' bodies. And so these dissections reinforced in him the belief that the male and the female must play very different roles in reproduction. Why else would they have such very different reproductive organs? The more important question was: what exactly was the difference between the male and the female role in reproduction?

One of the influential philosophies at the time was *atomism*, the idea that everything in the world is comprised of very small, indivisible, fundamental units – the intellectual birth of the atom. In terms of making babies, atomism was interpreted to mean that the male and female bodily fluids contained a miniature, perfectly formed version of the adult body of the respective sex, broken into parts, down to a pair of little arms and little legs, a compact torso, and a tiny head. When the male and female fluids mixed together during sex, these small parts simply assembled into a small body, which grew larger once it was sown in the fertile ground of a woman's body – the foetus. Conveniently, atomism explained how a child could resemble both mother and father, which made the concept quite popular among classical thinkers.

Aristotle did not agree with this atomistic view of the world. This was not, after all, what he saw happening in his experiments with birds. He had observed that hens would mate with more than one rooster. Yet, 'even when the hen is trodden by two males the offspring does not have two such parts, one from each male' – the only logical reproductive outcome, if you held to atomism. If the male bird supplied a miniature body part to each female with which it had sex, 'the offspring should have had a double portion', Aristotle argued, 'but it does not'. When it came to chickens and other birds, this meant the 'male supplies nothing material'. Likewise, of course, a woman who conceives after having sex with two men does not normally have a two-headed, four-limbed baby as a result. She isn't even very likely to have two babies, unless she happened to have twins. These were facts of life that Aristotle could also observe.

In *On the Generation of Animals*, Aristotle put forward an improvement in the reasoning for why there was a sexual division in reproduction, one that had nothing to do with the male and the female both providing the offspring's parts. In his scientific opinion, there were always two sexes in a species, because the male contributes the form and the female contributes the matter, the physical stuff of which the child would be made, or sculpted from. Form was superior to material. The male semen dictated the shape of the child, like a chisel gives a statue its shape, without itself becoming part of the product – the master artist at work. Since fathers created not just sons in their own image but daughters, too, daughters must, Aristotle believed, arise when the father's semen was weak. If the mother's reproductive fluid – her menstrual blood, in the philosopher's accounting – was also weak and could not be mastered by the semen, then you got neither a perfectly formed son nor a materially inferior daughter, but a monstrosity.

⊚⊙⊚

Aristotle's hypothesis may have been flawed, but it is not surprising that he did not consider a more accurate version of the inner workings of the female form – one was not available. Though Aristotle discussed the uterus in his book, very little had been revealed about the female reproductive organs at the time he wrote *On the Generation of Animals*.

The ovaries, referred to as 'female testicles', probably were discovered by an anatomist, Herophilos, who performed both animal and human dissections, some of them for public viewing, from his base in Alexandria, Egypt. But Herophilos was reportedly born in 335 BCE, just thirteen years before Aristotle's death. Soranus, a physician from an area of what is now Turkey, appears to have dissected human subjects as part of his investigation into obstetrics and women's diseases. He displayed a clear understanding of the various sections of the uterus, placenta, bladder, and vagina, which he described in great anatomical detail. Soranus's dissections, however, were conducted in the second century BCE – also well after Aristotle's time. For more than a millennium afterwards, little advance was made in understanding the true nature and function of these mysterious female parts, because in large part, human dissections were widely proscribed, which meant that cadavers were not openly available for this sort of poking and probing. Instead, physicians had to rely on the writings of Aelius Galen, the second-century Greek surgeon considered to be the most influential medical writer in all history.

Galen was born in Pergamon, the great cultural centre of Asia Minor under Roman rule. He came from a family of wealth and education, and he followed suit, training in philosophy, mathematics, and natural sciences. He had probably been influenced by his father in his choice of a career in medicine.

The story goes that the Greek god of healing, Asclepius himself, appeared to Galen's father in a dream to offer vocational guidance intended for his son. After this god-given training as a physician, Galen visited Alexandria, where the doctors placed great emphasis on the study of anatomy. On his return home, he was appointed physician to the gladiatorial games. This gave him the dubious privilege of regularly confronting the horrendous injuries inflicted in the arena. As ghastly as the job may have been, operating on the wounds allowed him to gain first-hand experience of human anatomy. He supplemented his observations of battered gladiators with dissections of abandoned corpses.

Galen lived some five hundred years after Aristotle, and medical knowledge had evolved. So he decided to develop his own theory of sex differences, based on his own work. In contrast to Aristotle's belief that the sperm was simply the seed that laid out the final form of the foetus, Galen thought the foetus's development was not just influenced but powered by the sperm, and that the female was actually a male in reverse. He was notably inspired by Herophilos, whose teachings were still popular in Alexandria and from whom he adopted the idea that a woman's ovaries were essentially testes. But Galen went further, positing that the female genitalia are identical to those of the male, only turned inward. According to this 'reversal' theory, the uterus was an inverted scrotum. This of course did not explain the function of those female parts that males lack – for example, more developed breasts. And the uterus did not serve the same purpose as the scrotum, a fact of biology that would have been understood even in Galen's day. But Galen was silent on these reproductive discrepancies.

When compared with modern views of reproductive evolution, though, Galen's reversal theory does not seem to have got everything wrong. For instance, in his essay 'Male Nipples and Clitoral Ripples', the renowned evolutionary biologist Stephen

Jay Gould argued that the man's body is not a basic structure from which a woman's diverged:

> Males and females are not separate entities, shaped independently by natural selection. Both sexes are variants upon a single ground plan... Male mammals have nipples because females need them – and the embryonic pathway to their development builds precursors in all mammalian foetuses, enlarging the breasts later in females but leaving them small (and without evident function) in males.

Likewise, Gould imagined that the clitoris and the penis were 'one and the same organ', their size determined by the relative balance of hormones, particularly testosterone, during foetal development. The same could be said of women's labia majora and men's scrotal sacs, though with these organs the presence of testosterone triggered a folding and fusing of the skin in the males. Gould took his argument a further step into controversy by stating that the clitoris was something like the appendix, an evolutionary artefact that no longer served a purpose. But on a more basic level, he supposed 'the external differences between male and female develop gradually', so much so that, 'from an early embryo so generalized that its sex cannot be easily determined'.

Since the early 1950s, when DNA was discovered to be the elusive matter that allows us to inherit traits from our parents, an incredible amount of scientific progress has been made. The complete genetic make-up, or genome, has now been mapped, or 'sequenced' in the jargon, for nearly two hundred organisms, including various kinds of bacteria and yeasts, honey bees, malarial mosquitoes, flies, worms, mice, rats, puffer fish, chickens, dogs, chimpanzees, and, of course, humans; our first draft of our genome was revealed in 2000, with the complete code cracked in 2003. This sequencing has provided a library of essential biological information. In addition, the various genome sequencing projects have determined that humans have around twenty-five thousand genes, divulged what some of these genes do, and confirmed chimpanzees as our closest kindred species.

DNA tells our cells what to do and when to do it. You can think of it as the genetic equivalent of an instruction manual for flat-packed furniture. It gets read, and the information it gives is translated into building a new piece of kit. The section of DNA that when read translates into the production of a certain chemical is a gene. Most genes are translated into a series of amino acids, and amino acids are the building blocks of proteins. Proteins, in turn, are the main constituents of cells, which collect into tissues, which themselves collect into organs.

Genes are made up of what are called nucleotides, which are molecules made up of sugars, phosphates, and chemical bases (referred to by the first letters, A, T, C, and G, of their chemical names). DNA is a long chain of these units of nucleotides, each built on one of the four bases. Geneticists refer to the chain by the sequence of individual bases of the nucleotides as they appear (for example, GATTACA, which is where the 1997 science-fiction film got its name). Not all sequences of letters 'spell out' genes; many just regulate genes, others seem to do nothing at all. Usually two chains of nucleotides wrap around each other – this is what gives DNA the double helix, or twisted ladder, look. These long strands of DNA double helices wind round in tight coils to form the chromosomes. Normal human cells have forty-six chromosomes, wound in two pairs of twenty-three.

Unlike Gould, Galen did not have the benefit of witnessing the minutiae of how human embryos develop, let alone knowledge of hormones or of DNA and chromosomes. So though the reversal theory might sound something like modern biology, Galen married his theory to assumptions about an unequal division of labour in the work of reproduction, assumptions that reflect the prejudices of the day. For example, among the ancient Greeks, another of the great differentiators between men and women was temperature. Around the fifth century BCE, a doctrine of health had formulated based on the balance in the body of heat and cold, dryness and moistness. It was widely believed that illness would erupt if one of these qualities dominated over another. Galen, like Aristotle before him, thought that women inherently had a different balance of heat and cold than did men. He started with the principle that women were colder, a state that influenced their behaviour and contributed to an inferior physiology and limited reproductive power. He even compiled an 'empirical' work on bodily heat, called *De Temperamentis*. And it was empirical: he had drawn his conclusions from experiments in which he had touched a range of different people – the old, youths, children, and infants – in order to uncover who were more and who were less hot. Throughout his report, Galen used the word *andres*, meaning 'men', to describe the participants in his trials, rather than *anthropoi*, meaning 'people'. That may be a distinction lost in translation, but it seems to indicate that the storied physician did not actually include any female subjects in an experiment from which he made the following judgements as to the nature of women:

> Within mankind the man is more perfect than the woman, and the reason for the perfection is his excess of heat, for heat is Nature's primary instrument. Hence in those animals that have less of it, her workmanship is necessarily more imperfect, and so it is no wonder that the female is less perfect than the male by as much as she is colder than he.

Though Galen did most certainly ponder, investigate, and experiment, there is no suggestion in any of his written accounts that his trials were ever conducted using women at all.

To Galen, the female of the species was not just inverted, she was incomplete, a view not substantially different from Aristotle's 'materially inferior' daughter. Consider Galen's analysis of the 'female testes'. As incomplete male testes, the ovaries should be expected to produce semen. But this 'female semen' would not be as pure, or as hot, as the male's, according to Galen. The ovaries, therefore, performed a function equivalent to the testes, but not as well. And he went 'one up' on Aristotle when he chose to refer to women as *arrostos*, a term most often used to mean a state of disease or morbid weakness. Unfinished, inverted, and in a state of morbid weakness – that's what women were made of.

The question was, if it was male semen that made babies, what did it contain that could hold such great life-granting power?

◎◎◎

At the time that Catherine de Medici was struggling to become pregnant, in the 1530s, most physicians still clung to such classical ideas of reproduction, by then more than two millennia old. And assumptions about the incredible potency of sperm animated the plans of the scientist later known as Paracelsus, who was studying medicine near Catherine's home city of Florence around the time of her birth.

Philippus Bombastus von Hohenheim – who styled himself as 'greater' than the Roman physician Celsus – spent much of his life formulating a recipe for the creation of human life. His recipe involved hermetically sealing a man's semen in a glass tube, burying the tube in horse manure for forty days, removing it, and then magnetizing it. Paracelsus believed that the entombed semen would begin to live and move, until it assembled into a miniature yet transparent human form, a *homunculus*, akin to the atomistic foetus imagined centuries earlier by the Greeks. After being unearthed, the homunculus was to be fed daily with *arcanum sanguinis hominis* – human blood – and constantly kept at the temperature of a mare's womb for a further forty weeks. From this protocol would emerge a human child, as normal as any child born of a woman, except perhaps a bit smaller.

In his own right, Paracelsus was a brilliant scientist, who made substantial and prescient contributions to the practice of medicine. Still, even in the sixteenth century, growing a baby in a bottle was mad-cap. So why did he think it plausible? By this time, many other notable scientists – from Galen of Pergamum to Leonardo da Vinci – had performed vivid experiments, including human dissections, to expose human anatomy. But many of Paracelsus's generation still found it incredibly difficult to cut the cord connecting their thinking to those of their forebears from the great intellectual centres of Greece. Though Paracelsus opposed many of the doctrines of the ancients, he espoused a definition of parenthood that would not be out of place in Aeschylus or Aristotle:

> The whole of the man's body is potentially contained in the semen, and the whole of the body of the mother is the soil in which the future man is made to ripen… [Woman] nourishes, develops and matures the seed without furnishing any seed herself. Man, although born of woman, is never derived from woman, but always from man.

Thus, horse manure stands in for the 'soil' of the womb, and a child is born.

Further, if a man's semen was believed to contain everything needed to create a mini-human, then any failure to become pregnant must be due to a fault in the incubation system – the woman or the horse manure, as the case may be. While Catherine de Medici applied scores of vile potions and lotions to her body in hopes of fertilizing the ground, Henry simply vouched for his virility by claiming that he had made another woman pregnant while away on one of his campaigns. To prove it, he went so far as to claim as a legitimate heir the baby girl of a woman who, according to some accounts, he had once raped (or at any rate, he had sex with on only one occasion). Catherine might as well have been born in ancient Greece, when women were not believed to be necessary for the production of children at all. Henry's omnipotent semen should have been more than enough. (She and Henry finally succeeded in their efforts a decade later, and went on to have ten children.)

◎◎◎

At the end of the sixteenth century scientists brought new tools to the question of the source of semen's power. In 1590, an early microscope was crafted by eyeglass makers in the Netherlands; within thirty-five years, Galileo Galilei had built his compound microscope, which he called his 'little eye'. Then, in 1670s Delft, a Dutch cloth merchant and surveyor named Antonie Leeuwenhoek turned his hand to lens grinding. Leeuwenhoek handcrafted around three hundred lenses, improving the technology from the poorer models that were available, though at first sight his efforts are barely recognizable today as microscopes. Crafted in brass or silver, he made them in a variety of tiny shapes; some looked like the flat end of an oar, others like an elegant handheld fan, a few like a toilet plunger.

Leeuwenhoek was more than a tinkerer, though, and used his microscopes to make a number of discoveries: of single-celled organisms, now called protists, in 1674, and of bacteria, two years later. He was also perhaps the first person to use these novel instruments to observe semen up close.

At first, it seems he was less than keen about putting semen under his microscope, or studying anything to do with sex, for that matter. This changed in 1677, when Johan Ham, a medical student, called on Leeuwenhoek at his home and presented him with a sample of semen that had been extracted from a patient with gonorrhoea. Ham thought he had seen small animals with tails writhing around in the fluid, and wanted confirmation. The claim captured Leeuwenhoek's interest. He began observing his own semen – acquired, he stressed, 'not by sinfully defiling', but from natural conjugal coitus. Through his crude microscopes he confirmed that there were 'a multitude of animalcules, less than a millionth the size of a coarse grain of sand and with thin, undulating transparent tails'. Since he had been studying his own semen, the animals were unlikely to have been parasites or linked to gonorrhoea – in Leeuwenhoek's scientific opinion.

Nevertheless, based on his reports, the tiny, tadpole-like creatures came to be known as 'spermatic worms', from *sperma*, Greek for 'seed'. In 1700, they were included in a book on human parasitology, *An Account of the Breeding of Worms in Human Bodies*, by Nicolas Andry, an influential proponent of the idea that life was generated only by sperm. In 1820, when the modern name *spermatozoa* – adding the Greek *zoa* for 'living being' – was coined, sperm were still considered to be a sort of parasite. (Around that time, Richard Owen, Charles Darwin's contemporary and bête noire, even classified sperm into the group of parasitic worms called Entozoa.) It is understandable that what appeared to be a moving, living being should have been taken to be a symbiotic animal that infected the life-infused semen of males, but not the reproductive fluids of females.

Having seen sperm first-hand, and being unable to detect the presence of anything similar in women, Leeuwenhoek himself began to suspect that the female ovaries were 'useless ornaments'. He noted that male rabbits that were grey only ever produced other grey rabbits – evidence that semen provided the sole contribution to the creation of offspring. He considered it 'proof enabling me to maintain that the foetus proceeds only from the male… and that the female only serves to feed and develop it'. Leeuwenhoek further claimed that his semen sported complex anatomical structures – nerves, arteries, veins – though no one else was able to observe them. He made a point of emphasizing these features in his drawings, noting that in semen 'there may be as many parts as in the human body itself'.

In 1694, the Dutch mathematician and physicist Niklaas Hartsoeker built on Leeuwenhoek's work to describe what the preformed animalcules looked like. Hartsoeker, who worked with rooster sperm, claimed that it was he who in fact had first discovered the animalcules in sperm, not Leeuwenhoek. In any case, it was Hartsoeker who first made the animalcules tangible to those who had not seen them with their own eyes. In his *Essai de dioptrique*, on optical instruments, he published a drawing of the *homunculi*, or little people, who inhabited each sperm. Hartsoeker described the egg as 'no more than what is called the placenta', once again defining the female's function as nothing more than nurturing a foetus that had been formed from semen, now sperm, alone. But then, Hartsoeker hadn't actually seen the animalcules with his own eyes; he had simply imagined that they might look like tiny, perfectly formed children, complete in every detail. As the head of one sperm, he drew a child curled up in a foetal position; in the other two sperm, the heads are children sprawled out, seemingly asleep or in a state of suspended animation. Each sperm's tail dangles from the children's pates like a Victorian man's nightcap. In his musings, Hartsoeker went on to suppose, correctly, that a foetus growing in a womb would require the means for becoming physically attached, in some way, to its mother. This, he proposed, was the purpose of the tail of the sperm, which would subsequently develop into the umbilical cord.

Hartsoeker's drawings represented no more than fantastical speculation, but five years later, in 1699, a French aristocrat and astronomer named François de Plantades reported that he had seen exactly what Hartsoeker had predicted. Peering through his microscope, Plantades said he had spotted miniature human forms, tucked inside the heads of each sperm. Perhaps for reasons of professional etiquette (he served as secretary of the Montpellier Academy of Sciences), he published his findings under the pseudonym of Dalenpatius, with his paper appearing simultaneously in London, Edinburgh, and Amsterdam. Dalenpatius's claim, however, was nothing more than a hoax, an attempt by Plantades to ridicule those who believed in preformed, make-your-own humans and microscopic animalcules. If his goal was to bring the whole field into disrepute, he was grossly unsuccessful. The existence of strange and mysterious creatures in sperm gained new credibility, and the little sperm people became entrenched in popular belief for the next one hundred years.

In this way, even though scientists now had the tools to investigate the body and no longer had to rely on intuition, many swore they saw things that simply did not exist – and would point to the microscope as their proof. And so reproductive science continued to remain faithful to the ideas promulgated by Aristotle and Galen.

These ideas were repeated in the widely circulated *Aristotle's*

Masterpiece, a compendium of medieval medicine and folklore thought to have been written around 1680. (It is also known as *The Works of Aristotle*, though it was certainly not penned by the philosopher.) *Aristotle's Masterpiece* includes some excerpts from his work, as well as of the writings of Galen and the tenth-century Islamic physician Ibn Sina, who himself wrote a commentary on Aristotle's findings. The book includes descriptions of midwifery, female reproductive organs, and all things related to sex and embryos. Because of its sexual content, it was considered pornographic, so much so that it was banned – and remained banned in the United Kingdom until 1960. In the United States, however, *Aristotle's Masterpiece* was more accepted. Until the middle of the nineteenth century, it was the most commonly read medical text – despite the arrival of the new microscopes, dissection tables, and complex experimentation, which completely contradicted the book's depictions of the workings of reproduction.

◎◎◎

For nearly two millennia, sperm reigned supreme. Then, it was discovered that mammals also had eggs.

The year was 1827, and a German scientist, Karl Ernst von Baer, was investigating the reproductive tract of a bitch. It had of course long been obvious that birds and reptiles had eggs; these were in plain sight. By the seventeenth century, it was suspected that mammals might have them, too, although no one had been able to find one. Leeuwenhoek had searched for a mammalian egg with his increasingly sophisticated microscopes, but he had thrown off the hunt as a lost cause. Using a better microscope, however, von Baer had been able to distinguish a yellowish-white, point-like object within some structures, called follicles, that he had taken from a dog's ovaries.

Von Baer was curious, so he sliced open a follicle, used the tip of his knife to remove the pin-prick object, and placed it under his microscope. 'It is truly wonderful and surprising to be able to demonstrate to the eye, by so simple a procedure, a thing that has been sought so persistently and discussed ad nauseum in every textbook of physiology as insoluble', he later wrote of his momentous discovery. He was 'utterly astonished' to see the egg with his own eyes 'and so clearly that a blind man could hardly deny it'. But blind men there had been aplenty – including Leeuwenhoek.

To Leeuwenhoek, eggs existed so that the preformed embryos in sperm could be implanted in them. His stubbornness is all the more surprising when you consider that in addition to the discovery of sperm, the Dutchman is credited with the discovery of parthenogenesis, the development of the egg into a new individual being without fertilization by sperm. If you weren't too sure that eggs existed, as Leeuwenhoek said he wasn't, you might say that this process amounts to a female bearing offspring with no lasting input from a male – the equivalent of a virgin birth. And Leeuwenhoek was the first scientist to notice that female aphids had virgin births all the time.

An avid gardener, in the summer of 1695 he became somewhat concerned that the leaves of his gooseberry, cherry, and peach trees were damaged. At first he thought the mutilations were the work of ravenous ants, but on closer inspection, he spied aphids. Leeuwenhoek did with the aphids what he did best: he pulled out a microscope and, as had become his custom, he searched for the eggs of this new species. He found none. He then dissected what he guessed were the females. He found no eggs in them either. But he did find miniature, preformed aphids. The first specimen he dissected contained four young, and he removed as many as sixty from another.

This should have put an end to the idea that male semen, or sperm, was the sole instigator of new life. But there again, Leeuwenhoek had chanced upon an organism in which reproduction is by no means straightforward. The sexual tactics employed by female aphids are tricky and complex. Two hundred million years ago, the insects evolved a reproductive strategy that allows them to practise reproduction by parthenogenesis – in cycles. This means that female aphids do have eggs, and both the fertilized and unfertilized eggs of a female are capable of forming embryos. The small aphids that Leeuwenhoek observed when he cut open his female – those born live as a result of parthenogenesis – were exclusively female. What is more, a single female generated by parthenogenesis may contain three generations within her body: the numerous embryos of her unborn daughters and, within them, her granddaughters-to-be in the early stages of development. For aphids this amounts to a brilliant strategy for rapidly producing an immense population; a virgin female can, in theory, produce billions of offspring in a lifespan of roughly one month. Here was a stack of Russian dolls, miniature yet fully formed creatures in ever smaller packages, just waiting to be born – a perfect preformed embryo, but from a female.

Despite this finding, and von Baer's production of the elusive mammalian egg from a dog, the egg continued to be considered the lesser element of reproduction into the Victorian age. In 1849, Richard Owen, who had classified spermatozoa as parasites, delivered a talk at the Royal College of Surgeons entitled 'On Parthenogenesis, or The Successive Production of Procreating Individuals from a Single Ovum (Egg)'. Owen had coined the word 'parthenogenesis', yet he could not extricate from his mind the influence of sperm over the process. Instead of the potential of eggs to self-reproduce, his lecture expounded on the virtue of sperm. For him, a 'virgin birth' could only ever follow an original fertilization event – and fertilization required sperm. What he called 'spermatic virtue' was a power contained in sperm that could be divided equally among countless offspring. He told his audience that the development of an embryo by parthenogenesis differed from a normal fertilization involving sperm 'only in… non-essential particulars', by which he meant that the power of sperm was the absolute requirement.

In the late nineteenth century, Owen would come under fire for this explanation. His critics were formidable – among them Darwin and Darwin's 'bulldog' supporter, Thomas Henry Huxley. Huxley was professor of general natural history at London's Imperial College and had contributed substantial knowledge to the growing field of comparative anatomy and palaeontology. He levelled great criticism at Owen's science and methods; in return, Owen published cloaked insults about Huxley's own work. It's fair to say that their earlier friendship had dissolved by this time. A colleague of Huxley's even advised him to shoot Owen in a duel. The two scientists had running arguments on anatomy, aphids, and parthenogenesis. Huxley particularly attacked Owen's references to a non-descript spermatic virtue or 'force' that could be retained through generations of aphid females, calling such speculations 'ignorance writ large'. For his part, Darwin egged on Huxley to challenge Owen on this point.

Meanwhile, in Germany another distinguished professor of comparative anatomy, Karl Ernst von Siebold, also ridiculed the belief in all-powerful sperm. Siebold did not respond by producing more speculation, but by performing exhaustive experiments to investigate 'true parthenogenesis'. This was, in contrast to Owen's definition, the development of an egg that was perfectly capable of being fertilized by sperm but which had not been. For his test subjects, Siebold turned to aphids, bees, and moths. He knew the conviction that eggs must be exposed to spermatozoa before they can develop was very deeply rooted; he himself had once been a strong opponent of the existence of parthenogenesis.

In 1857, after years of study, Siebold published his findings on bees and Psyche and Solenobia moths; in a nod to Richard Owen, he entitled his text *On a True Parthenogenesis*. In his exacting observations, Siebold had noted that the unfertilized eggs of his moths produced female offspring, but that queen bees produced male drones through parthenogenesis and female offspring from eggs fertilized by sperm. Contrary to Owen's definition, it was clear that new organisms could develop solely from eggs. Siebold also uncovered that not only can eggs develop into fully formed animals quite without any fertilization event but also that parthenogenesis was by no means an exceptional occurrence, something peculiar to aphids. He made a point of countering the idea that 'development of the eggs can only take place under the influence of the male semen'. This age-old concept, he wrote, 'has suffered an unexpected blow'. Rather than being a result of some undefined force of questionable existence, parthenogenesis was an independent, fixed, orderly event.

But if eggs could develop on their own, as Siebold had proved, then what was the point of the male? Based on Siebold's work, Darwin made a remarkable conjecture: 'I have often speculated for amusement on the subject, but quite fruitlessly,' he wrote to his friend Huxley, 'But the other day I came to the conclusion that some day we shall have cases of young being produced from spermatozoa … without [an egg].' Darwin had a point: if eggs could independently generate life, why couldn't sperm do it, too? And if not little people, what was inside the sperm, and how were these seemingly living creatures made?

◎◎◎

In 1905, Jacques Loeb provided an answer. Loeb, a physiologist working in Germany, was busy trying to force unfertilized eggs to develop into embryos. Using alkaline or acid solutions, potassium, and salt, even ox blood and cane sugar, he triggered development in the unfertilized eggs of sea urchin, starfish, marine molluscs, and other creatures. For the first time in history, someone had managed to create new life in the laboratory with *no sperm at all.*

In working out what to substitute for sperm, Loeb realized he needed to find something that must have two effects on the egg. 'In the first place', he wrote, to 'cause… its development' and in the second to 'transmit… the paternal characters to the developing embryo'. For the marine species with which he was experimenting, the ability to cause the embryo to develop was enough. Baby sea urchins born in his lab would need no fathers from which to acquire paternal characteristics. The same could not be said if the subject were not sea urchins but humans.

The fluid praised as the essence of life by Aristotle and Galen (and the innumerable others who came before and after) is indeed remarkable. Human semen is a rich cocktail, a combination of sugars, salts, enzymes, vitamins, and minerals, including such truly essential ingredients as fructose, sorbitol, inositol, phosphorus, zinc, magnesium, calcium, potassium, ascorbic acid (vitamin C), and cobalamin (vitamin B12). As the ancient thinkers suspected, it is also the medium through which a father provides his set of instructions for making offspring. But unknown to these early natural philosophers, some part of semen – about five percent of what a man ejaculates – contains fifty million to two hundred million sperm. These cells are highly specialized, built to travel up to four millimetres a minute and to release chemicals that can target and penetrate the egg.

The creation of sperm begins inside the testes of a pubescent boy, when the solid cords that had transected these glands throughout his childhood begin opening up into tubes. The process carves a space at the cord's centre through which fluids will eventually be able to pass. These tubes will become contorted and so numerous and fine that in an adult male testicle, their collective length will measure as much as 350 metres, or more than one thousand feet. They will also become home to the stem cells that become sperm, called *spermatogonial stem cells*, or SSC. Stem cells are by definition immature, in that they are somewhat undecided as to their identity and therefore retain the ability to become something different – something more definitive, more specialized. At the start of a wonderfully efficient production line, these rounded cells are the first widgets in the manufacture of mature, tadpole-like sperm and, ultimately, are the basis of male fertility. Each sperm is moulded out of the contents of these stem cells, then conveyed into holding areas, a bit like reservoirs, which line the outer layers of the fine tubules that now populate the testes. Driven by the male sexual hormone, testosterone, developing sperm will move in waves of output along the belts of these tubes, which eventually spiral like a corkscrew, in towards the space at the tube's centre. In the space of roughly sixty-four days, they will be transformed from round, nondescript cells to fully fledged sperm with heads and tails; from being tucked away in inventory to positioning themselves in readiness for consumption – ejaculation.

If that ejaculation happens in the context of unprotected sex, it will only be possible for one out of the many millions of sperm to make it successfully into an egg. If one penetrates the egg, the egg will harden to prevent another from entering. That is, if any succeed in entering at all. Achieving fertilization is a formidable task, and requires sperm that are fit for purpose. In the process of making sperm from stem cells, many defects occur. Their heads may be too large or too small, tapering or shapeless. They may even have two heads instead of one. Some sperm are made with bent tails, or tails that are too thin, too long, or too short, broken, coiled, or altogether missing. Some sperm have been found with a combination of defects. Sperm with tail abnormalities will have little chance of swimming well enough to get anywhere near an egg. Those with head defects may be carrying damaged DNA, or an abnormal amount of DNA. They may also be unable to use their head to penetrate an egg properly, even if they did get close enough.

The environment in which sperm are made may have a lot to do with how well they are formed, or the quality of their genetic material. Low levels of testosterone, and of the minerals zinc and selenium, seem to be bad for sperm and male fertility. Lifestyle and other factors of biology can also affect the integrity of the genes carried by sperm: exposure to radiation, heat, cigarette smoke, airborne pollutants, or chemotherapy drugs; sexually transmitted infections; ageing; a high body mass index; and medical conditions such as insulin-dependent diabetes – all can degrade the quality of DNA.

But even for perfectly formed sperm, there are many obstacles to overcome. In order to reach the egg, they must survive the acidic atmosphere of the vagina and avoid getting trapped in the sticky mucus in the cervix, the gateway to the womb. If they make it through those hurdles, they will then have to navigate the narrow entrance into the cervix, dodge cells of the immune system that will try to target and destroy them, swim upwards against the current, and escape a final molecular process that will screen and eliminate the vast majority of whichever sperm have survived. Sperm will do all that for only one reason. On a scale that is about one thousand times smaller than a mustard seed, the head of the sperm carries the genetic instructions to start making a baby – an essential ingredient of sexual reproduction. In fact, somewhat as Aristotle suspected two millennia ago, what the semen delivers into the egg will contribute to the form of a resulting child – its looks and its general genetic make-up – but not the 'matter'. This is because, if and when a sperm makes contact with an egg, only its head penetrates, so that it can release its DNA-containing package into the awaiting receptacle. This DNA is its only contribution.

In contrast, at one hundred times larger than a sperm, the egg is mostly composed of a large amount of cytoplasm, or cellular fluid. Cytoplasm is a repository of miniature organs, such as mitochondria, which produce energy for the cell. An early embryo developing in the womb will need all of the egg's resources to grow, until its own cells are able to perform these functions itself. And the growing foetus will indeed take 'matter' from its mother to build itself – calcium from her teeth to build its bones; nutrients and oxygen from her blood. But this is where any resonances with Aristotle's intuitions end. Like sperm, the egg also has its own unique complement of DNA, its own set of instructions. The DNA that was carried inside the head of the successful sperm must join with that of the egg, if there is to be any chance of creating a new human being. Together, the male and female genes sculpt the body of the future child.

As we have seen, in a normal human cell DNA is packaged into forty-six separate chromosomes, but sperm and eggs contain only twenty-three. One of these twenty-three chromosomes is the sex chromosome, chromosome X or chromosome Y. While an egg can only ever carry an X chromosome, sperm may carry either an X or a Y. If a Y-carrying sperm makes it to the egg, their forty-six coiled chromosomes will have one X in one strand and one Y in the other, making a boy. If, instead, an X-carrying sperm fertilizes the egg, the resulting child will be XX, a girl. On a chromosomal level, mothers can never determine the sex of a child – no matter what King Henry II and his suffering spouse Catherine de Medici might have believed. The sex of a child simply comes down to whether sperm from the father is loaded with an X or a Y.

There may, however, be other means by which mothers can influence the sex of their children. It is still controversial, but there is some evidence that, even before an egg comes anywhere near sperm, it may already be programmed to 'prefer' only an X- or only a Y-carrying sperm. If this were true, then an egg selected to accept only sperm carrying a Y chromosome would not develop into a baby even if it were fertilized by an X-carrying sperm. This preference of an egg for a Y-chromosome-carrying sperm seems to be influenced by higher levels of the hormone testosterone in the egg's immediate environment. And even after fertilization happens, the development of males could also be promoted by higher levels of glucose in a mother's body. Experiments with mice show that mothers on a very high saturated fat diet have significantly more male offspring than those on a diet with restricted fat.

Whatever other secrets the human egg holds, it is clear that it is more than just a passive recipient of semen. In fact, the ability of the eggs of insects and sea urchins to create new life entirely without sperm is no isolated occurrence. The more we look, it seems, the more we are finding that virgin births are happening throughout the animal kingdom, sometimes in the most unexpected places.

2

THE STORY OF SAFE SEX

I am particularly glad that you are ruminating on the act of fertilisation: it has long seemed to me the most wonderful & curious of physiological problems.

Charles Darwin, letter to T. H. Huxley, November 1857

We all know how sex works, right? If you're a young couple in your twenties (or younger), and you're planning to have a baby, or you had a baby around that age, you probably did not think too much about the process. If you're in your thirties or older, a bit more organization might be required: 'romantic' evenings in, planned around an ovulation predictor, possibly with the help of fertility drugs. You have 'unprotected' sex with your partner – throwing aside birth control pills, condoms, and other barriers to fertilization. Then you wait to see what happens.

You wait two weeks, maybe a month, and a period never arrives. So you visit your local pharmacist and pick up a pregnancy test, a urine-test strip that can detect the 'pregnancy hormone' *human chorionic gonadotropin*, or hCG. If hCG is present in a woman's urine, a sperm has entered an egg, and that fertilized egg has most likely made its way to the womb, where it is secreting hCG. If all proceeds well, twelve weeks after the icon appears on your pregnancy test, an ultrasound scan will present you with the image of a miniature human, about the size of your little finger. Can you tell who it looks like yet? Perhaps later, at the twenty-week scan, you will detect some familiar features; or if not then, when he or she is born. Whose eyes, skin, hair, and facial structure does this child have? After all, your baby will have DNA from both you and your partner, because sex was invented to mix up the genetic information within a species.

To find out why this is so, we need to travel back in time many millions of years.

◎◎◎

Sex first turned up around eight hundred and fifty million years ago, just as life as we know it made a leap from simple, single-celled bacteria. A new kind of cell was formed.

Called the eukaryote, it would be the common ancestor of all plants, fungi, and animals. The eukaryotic cell showed off a clever system of internal membranes, which organized and compartmentalized the cell, as well as a number of miniature organs, even an internal skeleton. One of its internal membranes surrounded a supremely novel creation – the nucleus, within which was contained the cell's DNA, the code of biological information that gave the cell its life; the DNA was coiled in chromosomes, packed there by proteins – zipped up, so to speak. That allowed the cell to combine DNA from two sources: the parents.

Prokaryotic bacteria, by contrast, have an external skeleton and free-floating, circular DNA. For bacteria, the problem of bringing DNA from two different ancestors together in a single cell – even a cell that is one-tenth the size of a eukaryote, as is typical – was solved in a variety of ways. Two bacteria cells could transfer individual molecules of DNA or small fragments of another genome back and forth, ostensibly by absorbing this stuff into their bodies. With the eukaryotic cell, whose DNA was neatly packaged up inside a nucleus, such fast and loose sharing could not work.

Eukaryotes instead evolved a method by which different cells could fuse. That meant combinations of whole genomes – not just individual DNA molecules – from different cells could be brought together, paired up, broken up, shuffled, and re-joined to make one new genome, which contained more genetic variety than either of the cells on their own.

Of course, recombining DNA was something that bacteria had been doing for ages – the machinery for the process had actually existed about three billion years earlier, during or even before the very first cell came into being. Long before sex was a twinkle in evolution's eye, the prokaryotes were using some of the tricks that sex-loving eukaryotes adopted, recombining foreign DNA into their own, most likely to grab spare parts that could be used to repair damage to their own DNA – a very different goal to generating genetic diversity or evolutionary novelty. For early eukaryotes, sex was 'selected' for its fidelity, for its ability to provide an accurate reproduction of the fused cells rather than random change. Sex preserved the innovations that set the eukaryotes apart from the bacteria.

Humans are eukaryotes, as are all animals, plants, and fungi, and we have selected sex. The human genome is made up of 2.9 billion DNA base pairs, over 700 megabytes worth of data, a lot to pack into a cell. Most human cells are about ten thousand times smaller than the fully extended length of our shortest chromosome, which, if fully stretched, would measure between 1.7 and 8.5 centimetres (about 1.5 to 3.5 inches). In order to carry around two metres (about six and a half feet) of genetic material, DNA must be highly condensed and stuffed and twisted in.

When each of us makes sex cells, that is, eggs or sperm, our twenty-two pairs of chromosomes and our pair of sex chromosomes – the XY of males and the XX of females – each duplicate themselves and line up in matching pairs. Each chromosome of a pair physically connects with the other at certain places along its length to swap genetic information, like dancers circling to and fro, touching hands and retreating back to the line, as in one of Jane Austen's house balls. When females make eggs, their chromosomes connect more often and at more places which means that eggs go through a more thorough shuffle of a woman's genes than sperms experience when males make sperm. The information that is swapped encodes the same sort of instructions; it's just that these instructions may vary in detail. That is how the genome becomes peppered with variations in genes, and how individuals may have different forms of the same gene, called alleles, at specific chromosome locations.

For example, there are a number of genes that shape, if not determine, the colour of your skin and hair. One of them, the melanocortin 1 receptor (*mcr-1*) gene, heavily influences your skin colouring and your potential to tan. The most common version of *mcr-1* allows immature yellow and red pigment molecules to be chemically altered to become brown and black. If you carry two copies of this common version you will be able to tan (as a bonus, you won't be as susceptible to skin cancers). But there are three other variants of *mcr-1*, which geneticists call *r151c*, *r160w*, and *d294h*; these variants block the transformation from yellow and red to brown and black. If you inherit one of these variants from one of your parents and the more common version from the other, you will be able to get a moderate tan. Inheriting a less common variant from both of your parents is likely to put you at an increased risk of skin cancer.

Your parents' chromosomes that carry these genes got chopped and recombined when they made the eggs and sperm that created you or other offspring. So the instructions that these eggs and sperm end up carrying is something of a mish-mash of the instructions that your parents actually inherited. That is to say, the way in which chromosomes are divvied up creates eggs and sperm (or an embryo, should they combine) with that unique mixture of parental genes. The lining up of chromosomes, the recombining of matching DNA from two individuals, ensures that genes with minor variations are mixed around any given population – although for much of human history, and indeed for many communities today, the pool of possible reproductive partners is rather limited.

On an evolutionary time scale, however, all this mixing is pretty insignificant. Major evolutionary innovations, such as the formation of a new species, or even the creation of an animal that is able to reproduce without sex, depend on the appearance of random mutations. In order for the early eukaryotes to preserve their exciting new method for recombining DNA – sex – mistakes and damage to their DNA needed to be picked up and corrected.

Sex was born not for promoting change and diversity but for limiting them.

◎◎◎

There is no doubt about it: sex is popular. The vast majority of species – 99.9 percent of higher animal species and about 92 percent of higher plants – reproduce sexually, at least on an occasional basis. A tremendous amount of research has been devoted to asking why this is the case.

The magnitude of the question becomes apparent if you

consider that, for females especially, having sex is dangerous, expensive, and foolhardy. The act of mating can be harmful, with sexual partners sustaining physical injuries or contracting sexually transmitted diseases. Finding a suitable partner involves an investment of time and energy – the work of trying to dodge those physical dangers as well as the risk of picking a mate that, in the worst-case scenario, never produces a viable offspring.

Even if a woman meets her 'soul mate' (a very human way of describing the safety of a reproductive partner), mixing genes may be a disaster; sexual reproduction breaks apart combinations of genes that work, after all, and they may not work in their new formation.

Moreover, because a sexually active female allows her mate's foreign genes to enter her body (and thus her offspring), her own genetic contribution is diluted by one-half. That seems like basic biology, but not all genes are created 'equal'. Most mutations, those mistakes in the combined DNA, arise in sperm.

Recent estimates show that the rate of mutations in males compared to females is two times higher in rodents, six times higher in most primates, and ten times higher in our primate, humans – which makes you question any concept that human evolution is the ultimate step in some hierarchical process. This comes down to a basic fact regarding how cells are made. Most mutations arise from mistakes in copying the DNA in one cell when it divides to make two cells, and they divide to make four cells, and so on. In cells that divide more often than others, more mistakes are likely to happen, simply because more copying is going on. It's a bit like a game of Chinese Whispers – the more times a message is passed around a circle of people, the more likely it is to get distorted. In men, cells divide to make ninety thousand sperm every *minute* – room for a lot of mistakes. And in the past century, geneticists have found that most of these mutations that occur in copying are bad news, and that the mutations frequently interact with each other, with bad results. This means that the prolific production of sperm is more likely to pass on harmful mutations than is the relatively more modest supply of eggs – and, of course, sperm pass their mutations on to offspring produced sexually.

So why, then, do females reproduce sexually, since it's not optimal for them or their progeny? This long-standing puzzle in evolutionary biology is known as the paradox of sex.

One of the first to tackle an answer was the German biologist August Weismann, deemed by Ernst Mayr to be second only to Darwin among the pioneers of evolution. In 1885, Weismann delivered a lecture series on the 'Significance of Sexual Selection'. Weismann believed that the reason sex had evolved and was retained was because it provided the variation upon which natural selection could act. Consider his example:

> Let us take the case of an insect living among green leaves, and possessing a green colour as a protection against discovery by its enemies… Let us further suppose that the sudden extinction of its food plant compelled this species to seek another plant with a somewhat different shade of green. It is clear that such an insect would not be completely adapted to the new environment. It would therefore be compelled, metaphorically speaking, to endeavour to bring its colour into closer harmony with that of the new food plant, or else the increased chances of detection given to its enemies would lead to its slow and certain extinction. It is obvious that such a species would be altogether unable to produce the required adaptation, for ex hypothesi, its hereditary variations remain the same, one generation after another. If therefore the

> required shade of green was not previously present, as one of the original individual differences, it could not be produced at any time. If, however, we suppose that such a colour existed previously in certain individuals, it follows that those with other shades of green would be gradually exterminated, while the former alone will survive.

This idea that, through sex, species are better placed to adapt rapidly should environmental conditions change or become hostile has dominated the discussion of the evolution of sex for more than a century.

Yet, natural selection on its own doesn't entirely account for the invention of sex, since random mutations also play a role – for instance, if our insect found itself saddled with a mutation that just happened to make it the same shade of green as the new food plant. Mutations perturb the genetic blender of sex, sometimes to the detriment of an individual offspring but sometimes to its benefit. There have to be other plausible explanations as to why sex is the number one way to make babies despite its drawbacks.

One modern hypothesis emphasizes the comparatively efficient way in which sex rids offspring of harmful mutations. Because genes are reshuffled among individuals in each subsequent generation, fewer bad mutations accumulate in a line of descendants. Sex makes it possible to 'reverse' a deleterious mutation by mixing DNA with a mate's. These mutations do not need to be harmful in and of themselves; they may simply provide an easy target.

Which brings us to the second reason why having sex to reproduce may be better than going without: it's all about ecology. In a fluctuating environment, sexual reproduction offers a short-term advantage since the genetic variability produced through sex offers chances to be better able to adapt. Indeed, the most popular of the ecological theories is the Red Queen hypothesis, which focuses on the advantages that sex provides in thwarting the threat of parasites.

Parasites are smaller and shorter-lived than their hosts, and so in general also reproduce more frequently and accumulate mutations more quickly than their host organisms. No matter how well adapted the target species' immune system might be, or how quickly it can change itself to deflect a threat, the parasites change even faster. They do not want to be made homeless, after all. To fight off potential assaults from numerous parasites successfully, host species create, on the scale of evolutionary time, an array of different gene combinations that throw up barriers against parasites.

The most important genetic weapons against parasites are our *mhc* genes, which encode instructions for the major histocompatibility complex, which is responsible for how white blood cells – the foot soldiers of our immune system – interact with one another, with other cells in the body, and with foreign objects. It will come as little surprise that *mhc* genes are the most variable genes contained in the genome of vertebrate animals; the range of forms in which the genes can appear is quite spectacular. *Mhc* variants determine how well our immune systems recognize invaders; how susceptible we are to infectious and autoimmune diseases; and even how we respond to odours, including body odours, and therefore things like our mating preferences and our recognition of others as kin (with whom we generally wish to co-operate). *Mhc* genes also influence the outcome of a pregnancy.

In their job as part of the immune system, MHC molecules on the surface of cells latch on to foreign molecules – known as antigens – from viruses, bacteria, transplanted organs, tumours, and other 'pathogens' that are not supposed to be in the body.

The molecules present these invaders to a subset of white blood cells, called the T lymphocytes, which in turn initiate an appropriate immune response. T cells do everything from destroying infected or tumorous cells, to remembering previous infectious attacks in order to call up a quick response to a repeat invader, to turning on other T cells and immune-system responders.

Mhc genes fall into two classes tied to these various immune-system tasks. Virtually all cells have *mhc* class I genes, which mainly provide immune protection from internal pathogens – pathogens, like viruses and bacteria, that have already made their way inside our cells. A few specialized cells, such as the antigen-presenting B cells and macrophages, have *mhc* class II genes. These cells engulf offending parasites. The MHC class II molecules bind to proteins on parasites and present the proteins to cells, which digest them all up, destroying the threat. There is a rare MHC class II variant that may give a particular advantage when it comes to parasites. Called *supertype* 7, it has been shown in lemurs to help protect the body against multiple parasites at once, and it is presumed to have a similar role in other primates, including humans. So, the more shuffling there is in *mhc* genes, the more chances a species has to 'out-think' pathogen threats.

In as far as the Red Queen hypothesis presumes that hosts and parasites are engaged in an evolutionary arms race, with nimble parasites able to produce more generations (and change) more quickly, the *mhc* variants that provide more resistance to parasites will be more likely to spread through a population. To stay ahead of the parasites, *mhc* genes will diversify, creating new combinations and rare types, like the *supertype* 7. The more variation, the better. And swapping genes through sex is a tested way of creating new combinations and rare types, and thus provides greater protection from a greater range of environmental pathogens.

Thwarting parasitic infection has knock-on effects when it comes to sexual behaviour itself. Among the males of some animal species, those individuals less infected with parasites typically have more energy to allocate to attracting mates. A landmark study, published in 1982 by British naturalist W. D. Hamilton and American evolutionary ecologist Marlene Zuk in the journal *Science*, investigated this question by looking at blood parasites infecting songbirds. Hamilton and Zuk looked at seven surveys of bird parasites, including several kinds of protozoa and one nematode worm, and found that there was a significant correlation between chronic blood infections and the striking displays that scientists associate with bird mating: bright male plumage, bright female plumage, and 'bright' male song. As they looked deeper, they also discovered that female birds scrutinized mates based on these displays, and they chose males that were marked with health and vigour, that is, those that were more genetically resistant to disease. This finding came to be known as the 'bright male hypothesis'. Perhaps the most spectacular avian plumage is the peacock's, and the peacock has often been used as an exemplar of evolutionary selection at work. Of course, when it comes to a peahen's mate, selection of a mate is relative. At the most basic level, healthy males are able to invest more bodily resources into maintaining healthy, vibrant plumes and engaging in elaborate courtship songs, dances, and other displays. Females may also tend to prefer less parasitized males in order to reduce their (and their babies') risk of infection, and male attractiveness might offer a clue to a potential mate's health.

Reproductive success has a genetic component in humans too. Men who carry a mutation in a gene on chromosome 7 that is linked with the gene for cystic fibrosis (the *cftr* gene) often experience infertility, sometimes because the vas deferens, a tube that should carry sperm towards the ejaculatory duct, may be absent or unable to provide the specific secretions needed for sperm to ready themselves for fertilization; and women with this mutation may have heavy mucus that prevents sperm from reaching the egg. A person with the genetic mutation may not have any other symptoms of cystic fibrosis, and may not be aware of the mutation until he or she has difficulty while trying to have a baby. The mutation causes a single change in a protein and seems to have benefits, however: the mutant CFTR protein is resistant to *Salmonella typhi* bacteria, the cause of typhoid fever. That change is more widespread among populations in Europe and Asia – in fact, one in twenty-five people of white European descent are carriers, meaning they have one of the two genes on that chromosome necessary for a person to have cystic fibrosis. So while the mutation can have serious consequences for a person's health if it is inherited from both parents, it appears to be the product of positive natural selection. So through sex, the genes that increase fertility, and which give parents' immune systems an enhanced ability to fend off parasites, are passed down to their children.

This isn't true only of songbirds and 'higher' mammals, and that includes us. Female green stinkbugs, for example, also choose mates on the basis of the males' potential genetic contributions to their young. These female bugs have a trickier time of it than peahens do, though, because large male stinkbugs, which are more likely to dominate energy resources and thus be more attractive to females than small males, are also more likely to suffer from parasitic infections. Even worse, it appears that body size is inherited, but only from the father – so there is more evolutionary pressure to prefer large males; larger offspring are more likely to survive against small bugs when resources are scarce. Being able to identify healthy large mates thus gives female stinkbugs a double advantage. Not only does the number of eggs they produce increase, but their male offspring have more success at attracting females and mating, a skill that seems to be inherited from the father, which potentially increases the number of descendants per 'son'.

The Red Queen hypothesis assumes that only the healthiest (that is, parasite-free) creatures are able to reproduce. These healthy creatures pass on their DNA to produce a genetic range among their young, which also have a better chance of avoiding parasite infestation. This evolutionary tactic does not strictly favour sex above no sex; rather, it favours diversity, however generated, over no diversity. Even in animals that reproduce without sex, genes are shuffled to a certain degree, but in relative terms, sex does more to mix things up. Not being able to effect or incorporate change into your genome is, in this view, a one-way ticket to extinction.

For this reason, it is surprising that there is little or no direct evidence showing that accumulating mutations or a reduced ability to adapt in the face of environmental pressures causes increased extinction risk in animals that only reproduce without sex. In fact, asexual species do not just die out as the predictions say they should. Those that survive use a range of unusual biological tactics to alleviate the negative genetic effects of their chosen approach to reproduction. These include being able to disperse their offspring to wide geographical distributions (large, sometimes very diverse habitats) and dormant resting stages (periods with low activity, sexual and otherwise) – mechanisms of maintaining population equilibrium that help to increase the chances of long-term survival for an animal with low genetic diversity. These tactics take the place of natural selection through sex, which weeds out the genes that are least adapted to the environment.

Sex provides survival strategies, but it is by no means perfect. Among other things, sex does not allow the creation of a clone from a genetically super-successful parent, a parent that has the ideal make-up for meeting a particularly harsh environment. This, however, is a problem that can be circumvented by approximating asexual reproduction, say, through inbreeding.

◎⊙◎

What makes a baby healthy and bonny, that is, a 'good' baby? In 1938, the eminent biologist J. B. S. Haldane wrote *Heredity and Politics*, a book he described as being 'addressed to such as are unacquainted with the science of genetics, but who are attracted or disturbed by eugenic doctrines'. The doctrines he discussed are disturbing indeed. Written at a time when, in many places, miscegenation was outlawed and apartheid was actively enforced, the book raised several controversial issues surrounding the genetics of human offspring.

Haldane asked whether inequality among men was fundamental and genetic, if the sterilization of genetic 'defectives' was appropriate or wrong, and what could be expected if mixed race children were accepted (and more regularly born into the world). He looked at whether certain races and certain social classes might be endowed with innate superiority, or stand as a 'pure' race – a belief he attacked. He wrote that recent learning about human inheritance had 'been used to support proposals for very drastic changes in the structure of society' – a clear reference to the treatment of Jews in contemporary Germany. He continued: 'And the stringent measures which have been taken… are said to be based on biological facts. I do not believe that our present knowledge of human heredity justifies such steps.'

After seventy-five years, the questions Haldane posed remain incendiary, and though our present knowledge of human heredity is growing exponentially each year, the sum total of accumulated data that has anything to do with race is minimal.

In 2008, two members of Parliament called for a ban on marriages between first cousins in the UK. In large part, their reasoning was based on data suggesting that Pakistani families from the West Midlands of England accounted for about thirty-three percent of the recessive genetic disorders in the region, but only around 4.1 percent of total live births, a dramatic statistic. The disorders were recorded, in the language of medicine, as 'recessive metabolic errors'. These are mistakes that only affect a child if that child inherits a copy of the 'bad', mutated, disorder-creating gene from both of his or her parents. And this is more likely to happen among parents who are closely related – who are consanguineous, that is, 'share blood'.

For several reasons, however, the statistics that the politicians used were skewed. There were problems with the way the data were gathered, and other studies over the years have shown that the risk of first cousins having a child with a recessive disease is quite low – no more than for the community overall. The issue wasn't that first cousins were marrying, it was that everyone in this particular community was slightly more likely than the general population to carry the mutant gene.

When epidemiologists want to investigate the chances that a certain group may be prone to recessive errors, they look at *genetic load*, the overall number of harmful mutations the average person is carrying, rather than specific gene mutations. Unfortunately, genetic load is not always simple to translate into morbidity and mortality rates – the chance that a foetus will not make it to full term, or that a child will have difficulty surviving into adulthood, let alone inheriting a genetic disorder. The medical statistics, for example, do not indicate that rare recessive genes are more likely to cause miscarriage than other factors. Moreover, imagine that one of the shared grandparents of two married cousins carried a gene for albinism, which would give each of the cousins' offspring a fifty percent chance of being albino. That doesn't necessarily mean that this grandparent carried any other recessive mutant gene, or that the children will automatically be unhealthy or have any other disorder – the genetic load would be small, but the chance of being albino would be relatively high. Finally, consanguineous marriages tend to be more prevalent among people with lower socio-economic status, which coincides, in and of itself, with higher rates of morbidity and mortality, as shown in Sir Michael Marmot's famous study of Whitehall bureaucrats. The politicians' motion was based on an oversimplified view of heredity and epidemiology, to say the least.

Apart from the fact that the proposed law did not take into account all of this evidence, it would have turned a blind eye to risky reproductive behaviours that are currently accepted among many other groups. It is not questioned, for instance, that women nearing menopause should be able to procure fertility treatments, even though it is understood that older mothers are more likely to give birth to children with chromosomal abnormalities; nor that people with Huntington's disease or other debilitating genetic disorders should retain the right to have children, despite an established fifty percent risk of the condition being inherited. Should a consenting adult be penalized for choosing a partner who might increase the risk of a genetic anomaly in his or her children? If so, would health services be required to screen potential reproductive partners, in the way that the charity Dor Yeshorim checks enrolled Orthodox Jewish families for a handful of recessive disorders before an arranged marriage goes forward? The law would have taken the choice of looking for a recessive trait out of the hands of the people having the child and put it in the hands of the government – a very drastic change in the structure of society. It would set a very disturbing precedent. Yet, the premise on which the law was proposed is quite basic: the premise that genetic variation is good and inbreeding is bad.

Inbreeding, of course, is what we normally call it when close relations mate. Between the closest of relations, we brand this *incest*, from the Latin for unchaste or impure. Genetically speaking, this labelling could not be further from the truth, however, because inbreeding *limits* the genes available to create any offspring, and so maintains a relatively pure familial gene pool; no truly foreign DNA is involved. As such, you could say that the ultimate form of inbreeding is making babies without a partner – that is, when reproduction does not involve sex at all.

◉⊙◎

The incest taboo is universal, though what it means for a particular community depends on whom a group defines as being too close a relation for sexual relations. In many Western cultures, there is an informal taboo around marriages between first cousins or between uncles and nieces. These proscriptions may be fostered partly by religious laws, economic imperatives, or long-standing prejudices, including the socially taught belief that children who grow up together cannot (or should not) develop a sexual attraction to each other. Regardless, most Westerners would not hesitate to say that consanguineous liaisons (those between 'blood' relations) are inherently *unhealthy*, triggering a range of physical and mental deformities – such as the out-sized 'Hapsburg jaw' you read about in school biology lessons, but also including infertility and early death. The conventional wisdom is so ingrained that, in recent decades, some scientists have begun to argue that the impetus to avoid inbreeding is itself genetic – that there is a gene that discourages inbreeding and promotes a taboo against incest in families that carry it.

In many cultures, however, consanguineous marriages, including between cousins, remain widespread. These marriages are most prevalent in Arab countries, with India, Japan, Brazil, and Israel following them in the rate tables. The liaisons are more common among people with less education (as well as lower socio-economic status), perhaps because higher status groups are more likely to have been influenced by Western beliefs about 'inbreeding'.

Conventionally, consanguineous marriages are considered to carry social benefits, such as being able to aggregate family wealth and ensure better treatment of the bride, and thus increase stability and security for the whole family. Many arranged marriages occur between 'blood' relations. But while those social benefits may hold sway in many families, there can also be biological benefits to marrying within the family: marrying a close relative might save your lineage's genes from extinction. In fact, there are cases in which inbreeding has actually facilitated a population's adaptation to an inhospitable environment and parasite threats.

Take, for example, the telling statistic that in many parts of Asia, the Middle East, and Africa, marriages between close biological relatives account for up to sixty percent of all unions. These geographic areas share a long history of exposure to malaria. Indeed, the five hundred million to eight hundred million people who are married to a 'blood' relation mostly live in the world's malarial regions. There are simply more consanguineous marriages in places where there is more *Plasmodium falciparum*, the protozoan parasite that makes its way into humans through the bite of the *Anopheles gambiae* mosquito and causes the most lethal form of malaria.

In 1949, Professor Haldane noted an apparent connection between malaria and a high prevalence of thalassaemia, an inherited blood disorder. Thalassaemia gets its name from the

Greek for 'sea' (*thalassa*) and 'blood' (*haema*), since one type of the disease is especially prevalent in regions circling the Mediterranean. The condition causes the body to make fewer healthy red blood cells and less haemoglobin, the iron-rich protein in red blood cells that carries oxygen around the body. In people with thalassaemia, this undersupply of red blood cells and haemoglobin leads to an undersupply of oxygen in the bloodstream. In its mild form, thalassaemia may cause tiredness, but when severe the spleen becomes enlarged, and the person may suffer from liver, heart, and bone ailments. Haldane believed that, for some reason, malaria seemed to be ameliorated among patients with thalassaemia. Could it be that the presence of malaria parasites in the environment had promoted the survival of the inherited genes that caused the disorder?

In humans today, the gene that causes the α+ type of thalassaemia is the single most common disorder caused by a mutation on a single gene, what is known as a monogenic disorder. Over five hundred million people carry the gene, and they live primarily in regions where malaria is or was endemic (in the past, malaria was endemic around all of the Mediterranean, including much of Southern Europe). As the Red Queen hypothesis would predict, when malaria emerged ten thousand years ago, humans adapted to the threat. In regions where malaria parasites were present, the population has much higher rates of gene variants – and not just those for inheriting thalassaemia – that decrease the likelihood that you will die from malaria if you are infected.

The variants that protect against malaria tend to change the structure or function of red blood cells. Among them are the genes responsible for sickle-cell anaemia (which changes the shape of red blood cells); the abnormal haemoglobin C (found mostly among Yoruba populations in West Africa); and the Duffy antigen negative blood group (which affects the protein on the blood cells on to which *Plasmodium* parasites attach). Having the gene for α+ thalassaemia, for instance, does not stop you from being infected with malaria, or from developing the symptoms of the disease. But it does reduce the risk of developing severe malaria, especially malarial anaemia, and therefore the risk of dying. (Although no one is as yet exactly sure why or how, people with thalassaemia also appear to be protected against developing severe anaemia, with or without malaria, and have greater resistance to lower respiratory tract infections.)

The practice of consanguineous marriages appears to map to those geographical regions most hit by malaria, and while that might be due to an assortment of factors, one of them is likely to be because inbreeding accelerates the selection of the α+-thalassaemia gene. Indeed, people with two copies of the protective gene variant have better survival odds against malaria than those with one or no copies. And, of course, to inherit two copies of the gene, both of your parents had to be carrying at least one, and there is more chance of this if they are related, even distantly. On average, twenty percent more of the children from consanguineous unions would survive malaria as compared to those from marriages between unrelated people. Inbreeding can sometimes be good for your health.

The usefulness of the adaptations that sex or inbreeding can provide assume that we must adapt to our environment to survive. For many humans living today, however, adaptation works the other way round. With the benefit of modern medicines and technology, we are increasingly able to adapt our environments to ourselves. Yet, in our globalized world, in which populations are mixing more than ever before, and offer less of the protections of inbreeding as a result, diseases such as tuberculosis and AIDS are among the leading causes of death. Both killers are caused by pathogens similar to *Plasmodium* in that they might be thwarted if a person possesses particular co-dominant or recessive genes – that is, by adopting the accelerated genetic resistance that comes from inbreeding. But sex is not designed to benefit the *individual*; it is designed to benefit the *population*. That is, in a nutshell, why it became such a popular strategy in nature. For most animals, the benefits to the species gained through sex are so great that they would only ever reproduce without sex under the most extreme conditions.

Men and women considering having a family will likely be thinking more about the individual benefits, costs, and consequences than the survival of the species. If humans had the option of reproducing without sex, would we do it that way? What sorts of extreme conditions might make us seek a virgin birth?

3

DESPERATELY SEEKING A VIRGIN BIRTH

Some people assert, and positively assert, that a female mouse by licking salt can become pregnant without the intervention of the male.

Aristotle, *Historia Animalium*, 350 BCE

Emmimarie Jones is thirty years old and completely absorbed in her domestic life, like any typically busy housewife. She has one daughter, eleven-year-old Monica, who is a happy English schoolgirl. Despite the utter normality of their daily experience, the mother and daughter seem somewhat strange to those who meet them. They share the same likes and dislikes regarding people, food, and clothes. And they share an uncanny physical resemblance.

Monica must have been conceived in the summer of 1944, when her mother was being treated for rheumatism in a women's hospital in Hanover, in Emmimarie's native Germany. Emmimarie recovered, but three months after she left the ward, her weakness returned. So she visited a doctor, intending to come away with a tonic to cure her ills. The consultation did not go as she had imagined. After examining her, the doctor said he was not surprised that she was feeling unusually tired – she would be, since she was pregnant. She smiled momentarily, in disbelief; she was sure he had made a stupid mistake. Emmimarie knew the facts of life, and she knew she had not been with a man. In fact, at the time she was supposed to have become pregnant, she was confined to the hospital, surrounded only by women – patients and staff.

Emmimarie insisted to the doctor that all she needed was a pick-me-up – some vitamins, perhaps, to help her feel a little less run-down. But the doctor was firm. He told Emmimarie that she would soon see that he was right.

Six months later, Emmimarie crawled out of the deep underground cellars where she had been sheltering from the Allies' bombing of Hanover. During the attack, her home had been destroyed and the city flattened for miles around, leaving a landscape of ruins. The draughty cellar would prove a difficult place to carry a child, but now she had nowhere else to go; it was her home. Her baby, Monica, spent the first two years of life underground.

After the war, Emmimarie married a Welsh soldier stationed in Germany with the Royal Engineers. When his service there ended, he took mother and child back with him to Britain. And it was in Britain that Emmimarie found herself embroiled in an international news exclusive.

The front-page headline of the *Sunday Pictorial* of 6 November 1955 shouted that virgin births were no myth, and a scientist could prove it. Find the Case, it commanded. Emmimarie could not believe her eyes. Here, finally, was a chance to uncover the truth about the circumstances of Monica's birth. She nervously put pen to paper, and wrote to the researcher quoted in the article:

Dear Madam

For ten year I have been wondering and worried about

the Birth of my Daughter. I honestly belief that she has no father. If you care to have all the facts please let me know. Before you write to me I must tell you that I am German in case you don't want anything to do with a German. I am married to a Welshman and have been in England over seven years.

Yours Sincerely
Emmimarie Jones

The letter was to reach the desk of a geneticist named Helen Spurway, also known as 'Britain's blonde biologist', the woman who had first grabbed the tabloid's attention.

⊚⊙⊚

Late in 1955, while working as a lecturer at London's University College, Spurway had found what she considered to be conclusive evidence that males were not necessary for making babies. Her conclusive evidence was that if you separate female guppies from males at the time they are born, the female fish still go on to reproduce. Further, the broods these virgin females hatch are unusual – they are almost entirely female. How could this be possible? Spurway wondered. There were only three likely explanations. One, that the mother fish somehow had contact with sperm while it was still an embryo in its own mother's womb. Two, that the 'female' was in fact a hermaphrodite – an animal like earthworms, snails, and indeed other species of fish that carry both eggs and sperm and could, therefore, self-fertilize. And three, that the fish were true virgins, and had no need of males for reproduction, in other words they underwent parthenogenesis.

Spurway knew that parthenogenesis was quite common in some insects, where the egg would start dividing inside the female without being fertilized, through some hormonal trigger. In the 1950s, scientists had even managed to force the eggs of cats and ferrets to develop into embryos all on their own, without sperm involved, so the process could conceivably occur in mammals. But these experiments had been highly artificial – the stuff of lab dishes, not of actual animals. Still, whether created inside a lab dish or an animal, a normal egg still only has one set of DNA. Whatever the species, this means that any offspring produced through parthenogenesis in a female could never have features that its mother did not. And that, Spurway thought, was the key to recognizing and proving a case of true virgin birth in a human.

Spurway decided to announce her guppy findings at a public talk entitled 'Virgin Births', presented at her university. At the end of her lecture, she mentioned her thoughts on using DNA to prove a virgin birth, said as something of a throwaway comment, a moment of speculation. But she did very openly suggest that there may actually be women out in the world who had given birth without having sex. It would be very rare, if it happened at all – otherwise, she noted, there should have been reports of fatherless pregnancies in women's prisons or other places of complete segregation. Perhaps, she proposed, there were women who suspected that they had experienced a virgin birth, but didn't mention it for fear of ridicule or social stigma. But if such women knew that their cases could be studied by scientists and doctors, even potentially verified, they might be more likely to come forward and speak about what had happened to them. Spurway added that odds were that a candidate child would be a girl and the spitting image of her mother. 'No faking would be possible,' she said. 'Blood grouping and skin grafting would give the proof.'

Spurway's words had been confined to the hallowed halls of University College and the ears of her fellow scientists, but her lecture had come to the attention of Audrey Whiting, an enterprising young journalist at the *Sunday Pictorial*, who had attended the event. After the talk, Whiting approached Spurway, asking for an interview, but the biologist waved her away, quipping that she did 'not speak to the popular press'. The journalist pushed ahead anyway with a report on Spurway's theory, and the editors put it on the front page.

The story, sporting the headline DOCTORS NOW SAY IT DOESN'T ALWAYS NEED A MAN TO MAKE A BABY, ran next to stiff-lipped assurances that the price of coal would not rise, advertisements for Brylcreem and Palmolive soap, and snapshots of demure starlets wrapped in demure bathing suits. Whiting informed her readers that there could be ten, maybe more, women in Britain who had given birth to a child without having sex with a man. But even for the tabloids, this was clearly outrageous, so Whiting had interviewed a number of doctors to offer opposing views. One issued a stern warning to the paper's readers, lest they got too carried away. 'Girls must not get silly ideas,' he said. 'The chance of a girl having a virgin birth would be twenty times less likely than her winning a football pool.'

Halfway down the page appeared the three words, in bold block capitals, that had stunned Emmimarie Jones – and doubled the newspaper's distribution for the day. Under the 'Find the Case' subheading, the *Pictorial* invited women to come forward, in confidence, if they believed their daughters were the result of a virgin birth. It stipulated that these women must be prepared to submit to examination by a panel of leading doctors, who were excited by the chance to identify a virgin birth in real life. To design the tests and interpret their results, advice was sought from the Medical Research Council, Britain's foremost authority in the area. If any woman's case was proved correct, she and her daughter were set to make medical – indeed, human – history. The tabloid's invitation ended with a quote from Helen Spurway lifted from the medical journal *The Lancet*: 'Remember, some of the unmarried mothers cited as curiosities by their contemporaries may well have been telling the truth.'

Remarkably, nineteen candidates came forward in response to the appeal. Of course, they included some of the 'innocent unmarried mothers cast out in disgrace by their families' that the paper had made sure to mention. Eleven of these women were immediately eliminated: they had thought that an intact hymen must indicate they had had a virgin birth, as for many societies, an intact hymen is a mark of virginity. But in some women the hymen can rupture spontaneously or through physical activity – playing vigorously as a child. And remnants of the hymen can persist in some women after vaginal intercourse, sometimes even after childbirth. Plus, the hymen is semi-permeable, so penetration is not required for fertilization (though in these situations, pregnancy is much harder). In any case, the presence of a hymen (or its remains) didn't mean that these candidates were truly virgins, and it certainly didn't mean that no sperm could have reached their eggs.

So the *Pictorial* published a more transparent explanation under the banner, YOU ASK WHAT EXACTLY IS A VIRGIN BIRTH? The newspaper's answer:

> Normally a man provides the seed which makes a child grow inside the mother. But in parthenogenesis (virgin birth) no man – and nothing from a man – is involved in any way at all. A virgin birth child need not be a woman's first child, and certainly need not be the child of a virgin.

After that, just eight candidates were left.

◉⊙◉

Of the eight women who were tested further, six didn't pass muster. The daughters had a different blood type from their mothers. Another mother–daughter pair was thrown out because their eye colour did not match.

Only one mother and daughter remained who passed all the preliminary tests: Emmimarie and Monica Jones. They also stood up to several more sophisticated trials. For example, they shared the ability to taste phenylthiourea, a chemical that has the unusual property of either tasting very bitter or being virtually tasteless, depending on the genetic make-up of the taster. Mother and daughter Jones were ushered into a consulting room and required to take a sip from each cocktail glass along a long row. Monica noted that the drink in several of the glasses tasted funny; Emmimarie thought so, too, and for the identical drinks. Monica said the experiment had been great fun; she reported to the *Pictorial* that the doctors' rooms looked like a bar.

Then they took a substance secretor test. Around eighty percent of people with a European ancestry come up with a positive result in this test, which looks at whether or not you have the so-called secretor factor, something like an honorary blood group. The genes that make you a secretor are found on chromosome 19, so the test was a crude way of determining whether Emmimarie and Monica had the same genes at that location. Being a non-secretor seems to have several disadvantages. It is associated with insulin resistance syndrome, where the body becomes less efficient at lowering blood sugar levels, and lowered levels of antibodies, which put you at greater risk of infection and illness. Non-secretors are especially prone to *Candida* organisms, such as the yeast that causes thrush, and tend to have more problems with heart valve disturbance as a result of infections after dental work. They may also be at increased risk of recurrent urinary tract infections and a variety of autoimmune diseases, including reactive arthritis, multiple sclerosis, and a condition called ankylosing spondylitis, which can lead to fusion of the spine. All of which shows how a small change of one gene on one chromosome could manifest itself as a big difference between a mother and a daughter.

Now, if you are a secretor, your blood type – the classic A, B, AB or O – will show up not just in your blood, but in your saliva, sweat, tears – and in your semen, if you are a man. If you are not a secretor, your type will not show up in these other fluids. Emmimarie and Monica were both blood type A, and the saliva of neither mother nor daughter contained substance-A secretions. So again, Monica was indeed the spitting image of her mother.

The final preliminary test looked at patterns in the blood-serum proteins of mother and child. These proteins were separated by size, and an image created that showed the line-up of the proteins, which could be compared. The Joneses were an identical match at every size.

But there was one final test that the pair was required to pass, should they be willing to take it. Helen Spurway saw it as the test that could provide the conclusive proof that Monica was fatherless, and had no genes other than those of her mother. Other scientists, however, believed its results would be obscure, at best. Nevertheless, absolute secrecy was assured, because if the doctors found out that Emmimarie Jones was not, in fact, a virgin mother, not one word would ever have been published to the world. The test in question was a skin graft.

When skin is grafted on to a body, the body's immune system will work to reject it as a foreign body, unless the donor is genetically similar to the recipient. This is why many graft surgeries involve taking donor skin from another site on the person's own body, known as an autograft, and why people who have undergone a graft from another person (who is not an identical twin) or another species must take immunosuppressant medications long after the surgery. The test Spurway proposed was to take a piece of Monica's skin and graft it on to Emmimarie's body. If the mother's body allowed this graft to persist indefinitely, without breaking it down, that would prove they were a genetic match – that there was nothing in Monica's skin that was considered to be 'alien' to Emmimarie's body. Spurway also realized that doing a graft the other way round, from mother to child, would not work; the mother would have antigens, substances that her immune system would be able to protect her against, which the child did not.

Emmimarie Jones readily agreed to the operation on herself and her daughter. Monica agreed as long as she could have lots of comic books. So, shortly before Easter 1956, Emmimarie and Monica left their English home, armed with adventure comics and destined for the secret location of their secret operation. Consultations began among the research team, blood specialists, and plastic surgeons, and the grafts were done both ways: Emmimarie was transplanted with her daughter's skin, while Monica wore her mother's.

◎◎◎

Through all of this battery of tests to find a virgin birth, Emmimarie Jones and young Monica were in good company. Theirs was simply the post-war, boom-time contribution to a long list of virgin mothers, from saviour gods to supernatural impregnations – or so the stories go. Almost always in these tales of virgin birth, the hand of God is involved, and there seems to be no culture that does not tell the tale.

The pantheon of gods is populated with virgin births, in heaven and on earth. The river nymph Nana was miraculously impregnated by a falling pomegranate, and her son Attis became the lover of Cybele, the mother of Greek gods (making Nana the grandmother of the gods). Hera, the wife of Zeus and thus the queen of heaven, renewed her virginity every year at the holy waters of Kanathos. She spurned the unfaithful Zeus, and all mortal men, to conceive her son Hephaestus. Zeus impregnated Leto, who bore the twin gods Apollo and Artemis, just two of the many children he sired by virgins. Artemis and her half-sister Athena were said to be virgin mothers too. Kausalya gave birth to the king Rama, the seventh avatar of the Hindu god Vishnu, after drinking nectar that had been made in offering to the gods. The virtuous Sita, Rama's wife, was the offspring of the land itself. Kunti of the epic Mahabharata was impregnated by the sun, Surya, when she recited a mantra that summoned him to her for that purpose. Queen Maya of Nepal, who hadn't become pregnant in twenty years of marriage, claimed she was spirited away in her sleep to a mystical lake, where a white elephant, holding a lotus in its trunk, circled her and then entered her womb, to later emerge on earth as the Buddha. The Aztec Coatlicue fell pregnant with Huitzilopochtli through the touch of a ball of feathers as she napped in a temple. In Babylonia, creation itself came about when a divine wind hovered over a female abyss called Tiamat; and Venus, the Roman goddess of love and fertility, has been perversely worshipped as a virgin.

Pre-dating them all was Isis, the sister and wife (and in some versions the mother) of Osiris, who was fabled in Egypt for having been deflowered in her own mother's womb, a bit like Helen Spurway thought her guppies may have been. In the land of the pharaohs, there was also the queen Mautmes, who was visited by the ibis-headed Thoth, the messenger of the gods, and informed that she would bear a son, though she was a virgin. Carved on the wall of the temple of Luxor, there are scenes depicting Mautmes as she is escorted by the holy spirit Kneph and the goddess Hathor to the *crux ansata*, the cross that symbolizes life, through which she could be impregnated with a touch of her lips. Setting aside the need to hold her mouth to the cross, this story might sound quite familiar to anyone who has heard the tale of Mary, mother of Jesus (who was also one of many virgin mothers with a form of that name, including Myrrah, the mother of Adonis, and Maia, mother of Hermes).

Among the ancient peoples circling the Mediterranean, the idea of a mystical birth probably gathered popularity through the veneration of a scroll about a virgin mother goddess based at Sais, an ancient Egyptian town in the western Nile Delta. The patron deity of the town was Neith, a goddess that the Greeks, including Herodotus and Plato, would later identify with Athena, since Neith, like Athena, was both the goddess of war and the goddess of weaving and the domestic arts. Because of this, Neith was the protector of women and a guardian of marriage. But her original role, dating to around 3000 BCE, was probably as a symbol of creation and fertility.

Neith was a goddess praised as a virginal mother and nurse, a mysterious mixture of virgin female and fertile mother that had great resonance among those who imagined it. So that many of the great men – the saviours, philosophers, and conquerors – were cemented into a demi-godly status with reports that they had come into the world over which they had power through such a birth. These virgin birth celebrities included Confucius, Plato, Alexander the Great, Julius Caesar, and Genghis Khan.

The legends of virgin birth are a counterpoint to the ancient notions of how regular babies were made. Over the thousands of years during which doctors and scientists said that women were just a vessel for carrying babies, not a contributor in creating them, nearly anything seemed capable of making a woman pregnant. Things like exposure to the sun or the wind (as recounted in myth) or to a fire. Or perhaps you ate pomegranates or magical fish or licked salt, like Aristotle's mouse. Or, you made a wish, stood under a shadow, happened upon a holy spot, or were breathed on by a god. These were all possible reasons why a woman might be pregnant, because no one understood then what was really going on.

In post-war Britain, however, there were few such illusions, and a virgin birth was, for the most part, held up as a rare, immaculate occasion, reserved for very special cases and very special storytelling. The *Sunday Pictorial* received complaints about the story's effect on younger readers, who it was believed were being exposed to far too many particulars about the mechanics of sex and pregnancy. To that, one of the doctors advising the newspaper retorted that any children old enough to read the tabloid *should* know about childbirth.

Most of the angry letters, of course, came from people worried that a virgin birth involving an ordinary human would 'undermine the character of Our Lady's virginal conception' and shake the foundational beliefs of people adhering to the Christian faith. The reaction from the Church was more deliberated. A Catholic publication, called *Universe*, carried a 450-word response, printed five days after the *Pictorial*'s front-page article, that drew on scientific evidence as well as matters of faith:

> Parthenogenesis, or virgin birth, is not entirely unknown in the economy of nature... Now if God could have endowed such creatures with life, and then bestowed upon them this otherwise unknown method of propagation, would it have been difficult

> for Him to bring about the birth of his only begotten son in a parthenogenetic manner? He who can do the one can just as easily perform the other wonder. If it were true, as Dr Spurway has confidently asserted, that one woman in a million might produce a child which never had a father, this would in no way undermine the miraculous character of Our Lord's conception and birth.

Quite so.

The Church's diplomatic handling of the issue was not an example followed in the scientific community, who were far less enthusiastic about the statistics. This remains a little surprising, when you consider that there was actually no plausible scientific explanation at that time for why women should not be able to reproduce without men. In the lab, after all, scientists had already succeeded in inducing pregnancy in rabbits without mating. The researchers had discovered that all it took was cooling down the Fallopian tubes, the tubes that connect the ovaries to the womb. Yet, a *Lancet* report that appeared around the time of the tabloid search declared, 'No "reasonable man" would even entertain the possibility that a woman might become pregnant without a single sperm entering her uterus.'

◎◎◎

Once Emmimarie's and Monica's blood samples were found to be a match, the test results were checked and double checked, then presented for debate. 'Doubting doctors,' as the *Pictorial* put it, 'who had in effect set themselves the task of breaking down the mother's story became less certain that they were on the winning side.' The paper went on to note that 'several of the medical men who had been sceptical about the outcome of the investigation now became keenly interested'. Based on the new evidence, they certainly should have been. If a daughter had a father, the likelihood that the battery of tests to which the Joneses had been subjected would have yielded such a clear agreement between mother and daughter was less than one in a hundred.

It was Helen Spurway's husband, Jack, who worked out the maths – he was the person who did the calculations to measure the similarity between Emmimarie's and Monica's blood. Jack was known to the world as J. B. S. Haldane, and the two eminent scientists had been wed nine years previous to the virgin birth investigation. At the time of their marriage, Spurway was one of Haldane's students, and twenty-five years his junior. He may have been one of the most prescient scientists of the twentieth century, but this founder of modern genetics was also famously colourful and eccentric, well known for experimenting on himself. In the name of medical research, for example, he once shut himself in a room full of carbon monoxide and swallowed bicarbonate of soda with hydrochloric acid, although not all at the same time.

This temperamental tendency was shared by husband and wife. As a couple, they loved to shock and argue – loudly, and especially in public. The pair often took their students out to the pub, to discuss work, politics, and people. On one occasion, Spurway drank three and a half pints of ale, staggering thoroughly drunk into the street, and straight on to the tail of a policeman's dog. The policeman remonstrated, at which she shouted, 'That's what dogs' tails are for!' – then punched the policeman in the stomach, adding, 'And that is what policemen's stomachs are for.' She was fined £20, but refused to pay it, choosing instead to be arrested and serve a two-week stretch in Her Majesty's Prison Holloway.

When it came to her science, however, Spurway was better behaved. Unlike her husband, she was no great theoretician, but she was a meticulous observer, and always committed to honesty about the facts. She took pains to stress to her students the absolute importance of writing down what they saw, not what they would have liked to see but what actually appeared. And this would be key to her final interpretation of Emmimarie and Monica Jones's remarkable results.

Despite the complete match between the Joneses' blood, there was a problem with the skin graft test. They were, apparently, incompatible with Spurway's hypothesis, and her expectations. The bit of Monica's skin grafted on to her mother had been shed in approximately four weeks, and the skin from Emmimarie grafted on to Monica had remained healthy for longer. It took six weeks before that graft from the mother began to lose its blood vessels, a sign that the skin would soon detach. In other words, Monica's skin contained something that Emmimarie's immune system did not recognize, while Emmimarie's skin did not offend Monica's system as badly. Was this a sign that Monica had DNA that her mother did not have? Was it a father's genes that caused the mother to reject her daughter's skin?

Eight months after the search for a virgin mother had been announced, the *Pictorial* published a world exclusive on Emmimarie and her daughter, relating their biographies and the battery of tests. For the serious medical reader, the full details were revealed in *The Lancet*, which published, 'Parthenogenesis in Human Beings' by Dr Stanley Balfour-Lynn of Queen Charlotte's Hospital in London. Balfour-Lynn, supported by a pantheon of distinguished doctors, had put the mother–daughter candidates through the necessary medical tests. On the point of skin grafts, *The Lancet* piece concluded that they indicated that Monica's genes did not in fact match her mother's, despite all evidence to the contrary. Emmimarie and Monica had failed the final and, in Spurway's expert opinion, the most conclusive of the compatibility tests.

But there was a curiosity planted in the centre of this scientific result. What any parthenogenetically conceived child certainly could not have, unless they had mutated, were any genes that had not come from the mother in the first place. This is why the skin graft from a virgin-born child would be expected to take when implanted on his or her mother, but one from the mother would not necessarily take on her virgin-born child. Yet, the opposite had happened in Emmimarie and Monica's test. What could be going on?

In such a case, Balfour-Lynn wrote in *The Lancet*, interpretation was difficult, making rigorous proof impossible. The one thing that was clear, however, was that Emmimarie must have believed what she claimed to be true. It was unlikely she would have set out to deceive people into accepting a virgin birth hoax, especially once she learned of the battery of medical tests that she and her daughter would have to go through. Yet, she happily agreed to run the full gauntlet. The medical journal compared Emmimarie's belief with cases in which the absence of 'pre-knowledge' has been taken by courts of law to constitute proof of the rightness to a claim. Unfortunately, the absence of pre-knowledge is not something that can be precisely evaluated by science. And so the final conclusion of the controversial study was that Emmimarie Jones's claim that her daughter was fatherless must be taken seriously, and that the doctors and scientists involved would have to admit that they had been unable to disprove it. The *Sunday Pictorial*'s triumphant interpretation of that verdict: 'Doctors have been unable to prove that any man took part in the creation of this child.'

While the tabloid version might be true in the most literal sense, there is no getting away from the issue that the study was in fact entirely inconclusive. The doctors' analyses of the Joneses was consistent with what would have been expected in a case of a female-only reproduction, except for the skin grafts. But did the fact that mother and daughter rejected each other's skin grafts mean that Monica was or was not the result of parthenogenesis? The only way to know for sure would be to get hold of some DNA from Emmimarie and Monica, and perhaps from Emmimarie's parents, because today, analysis of the subjects' DNA would reveal – or, at the very least, suggest – reasons for the intense similarities between the two Joneses.

It is safe to say that the odds are, overwhelmingly, that Emmimarie Jones was no virgin mother. She is also unlikely to have set out to deceive the scientists, which opens the possibility that she may well have been taken advantage of during the hospital stay during which she must have become pregnant. But it is also intriguing to consider that the tabloid-dishing scientists had found something extremely rare – something that would not be recorded again until forty years later, when a boy was identified who had his mother's blood, but not her skin.

◉◉◉

At the end of a report published in the October 1995 issue of the journal *Nature Genetics*, three photos capture a toddler boy identified only as FD. The centre portrait depicts a lovely cherub who could have easily fronted a promotional campaign for some wholesome baby food. To the right and the left, there are pictures of him in profile: the image of perfection from the right, but from the left, a puzzling confusion. The lower half of the boy's face is underdeveloped, out of sync with the rest of his body.

FD first came to his doctors' attention because of a blood test.

The test had come back with an unusual result: FD had two X chromosomes – for a boy, one X chromosome too many. So, even though he had testes and a penis, FD should have been a girl.

Strictly speaking, to be a boy you do not always need a whole Y chromosome. There are particular sections of the Y chromosome, notably one called *SRY*, that are essential to making a man a man. In cases like FD's, it is often found that these important sections of the Y chromosome attach to an X chromosome – just enough to make someone male. But FD showed no sign of having Y-chromosome genes on either of his X chromosomes. And yet, he was very clearly, at least when it came to his physique, a bouncing baby boy.

Next, FD's doctors decided to analyse skin from his right ankle to see if they could shed some light on his 'true' gender. In FD's skin, but not his blood, they found evidence of some Y chromosome material. So whereas his blood said that he was some kind of abnormal female, his skin said he was a genetically normal male. It wasn't so much that FD was both male and female but that the toddler was a chimaera: different parts of his body appeared to have been made from cells containing different DNA – that is, from different beings altogether.

The question was, how did the child get this way? One possible answer might be seen in the case of a woman, known in the literature only as 'Jane from Boston', who had needed a kidney transplant. Jane had three sons, all of whom were willing to donate a kidney to her, if they proved to be a suitable donor. And the likelihood was that, since children share half their DNA with their mother, at least one would turn out to be a good match. Jane had every reason to expect good news when the test results came in. Instead, she found herself opening an officious letter from the hospital informing her that two of her three sons were not actually her kin.

Since Jane had conceived all of her sons naturally with her

husband (who DNA tests showed was definitely their father), the results effectively suggested that she had somehow given birth to another woman's children. That, of course, she deemed an impossibility, especially since mistakenly swapping not one, but two, children at birth would have been a highly improbable coincidence. Further checks had to be done. Doctors tested DNA from other tissues, including Jane's thyroid gland, mouth, and hair. And that was how they discovered that this woman's body was composed of two genetically distinct groups of cells. Jane, like FD, appeared to be a mixture of two different people.

The most likely explanation was that Jane's own mother had conceived non-identical twin girls, who would have been no more alike genetically than two siblings conceived at separate times. At an early stage of the pregnancy, however, these twin embryos had fused, to form a single embryo. Because Jane's blood cells presumably carried DNA from one twin, and her ovaries and the majority of her eggs carried DNA from the other, a quick DNA test threw out the result that she was not the mother of two of her own children. Speaking in terms of DNA only, Jane's unborn twin was, in fact, the mother.

Yet, in one very significant way, FD was not like Jane at all.

◉◉◉

Getting pregnant in the normal way is a hit-or-miss process; a couple has only around an eighteen percent chance that the man's sperm will penetrate the woman's egg at her peak period of fertility, assuming, of course, that they are having unprotected sex. Then, there are the odds that a fertilized egg will transform into an embryo, and make it to full term. Jane's foetal life had been very rare – two different eggs developing into two different embryos and then fusing together. Which makes the way in which FD began life all the more incredible. FD had originated from only one egg. What is more, that egg had broken the laws of nature and developed into an embryo without waiting to be fertilized. His blood with its two X chromosomes was the product of parthenogenesis.

How tiny is the likelihood of each single event in the sequence of events that brought FD to life! First, one of his mother's eggs – the egg that would become him – was activated by some hormonal trigger, despite there being no sperm around to do the job. The activated egg then began dividing, the first steps towards becoming FD. The only DNA in FD's embryo at that point was from his mother. Next, rather 'miraculously', along came a sperm – a single sperm, as far as the scientists can tell – from his father. It should have arrived too late to have any effect, since normally after an egg is activated, a cascade of chemical signals tells the egg's outer layer to harden, making it impossible for 'follow-on' sperm to penetrate the egg and mess things up. An egg that accepts more than one sperm will form an embryo with too much DNA, and such embryos are normally destined for early termination – a miscarriage.

If FD was the product of parthenogenesis, couldn't Monica Jones have been a product of parthenogenesis, too? Unfortunately, scientists will probably never be able to answer definitively the question of Monica's maternity and paternity; they just don't have the information necessary to figure it out – the DNA of her mother's parents. But there might be alternative circumstances in which a woman might give birth to a child, apparently unaided.

Take, for instance, the best-known story of a virgin birth and consider how in the world such a miracle could have been effected – biologically rather than divinely speaking. In the third century CE, an influential Church father named Origen worked to promote belief in the virginity of Mary, mother of

Jesus, in what sounds at times like evolutionary terms, to the modern ear. The evangelist wrote:

> For it is ascertained that there is a certain female animal which has no intercourse with the male (as writers on animals say is the case with vultures), and that this animal, without sexual intercourse, preserves the succession of the race. What incredibility, therefore, is there in supposing that, if God wished to send a divine teacher to the human race, He caused Him to be born in some manner different from the common!

Origen argued that reproduction without a mortal man was very uncommon indeed; for instance, he dismissed the legend of his countryman Plato's immaculate conception, as 'veritable fables' that did not demand such creative innovation. But what rare biology might have been involved in producing Jesus (or for that matter, FD)? To answer that question, an emeritus professor of genetics at University College London, Sam Berry, has worked out the biological possibilities for how Mary could have given birth to a son, while still remaining a virgin. Beyond the issue of activating the egg to develop into an embryo without a human sperm, there is another problem: the fact that Jesus was not Mary's daughter.

A woman alone should never be able to provide the genes needed to make a son. The Y chromosome, which carries the genes that dictate maleness, are normally only carried by men, so are passed on through a father. So if Mary had given birth to God's divine daughter, the biology would make sense – she would have been able to provide all of the DNA, with the supernatural power activating the egg in some way. But Jesus was, as we know, God's only son.

To produce a son via virgin birth, Berry suggests that Mary

may have suffered from one of the several chromosome abnormalities that cause testicular feminization, a condition which affects around one in thirteen thousand people. While women with testicular feminization have a Y chromosome, they 'present' as a girl at birth – with a vagina and no testes or penis. And throughout life, they appear to develop along female lines, growing breasts, for instance. Internally, however, their bodies tell a different story. The vagina is quite short, and leads to nowhere – neither to womb nor ovaries – and hidden away in the abdomen, there is a set of testes. Normally, testes make testosterone, the hormone that masculinizes the growing embryo and child, so that he develops the external genitals and other sexual characteristics of a typical male – such as more extensive body hair. Yet, even though a woman with this condition is exposed to the testosterone produced by her testes, the body seems to be insensitive to its effects. As a result, she will likely show off a luxuriant head of hair and never experience male-pattern balding – and further, she'll never develop hair where you would expect it, in the armpit and pubic area.

Essentially, Berry's idea was that the only way a woman could have a son without input from a father is if she herself carried a Y chromosome. In fact, a son from a mother who carried a Y chromosome – most likely one that was not fully functional – might develop into a normal male, without issues of abnormal testosterone production or sensitivity, if, say, that chromosome mutated back into a functional state somewhere along the way.

But what made this biologically plausible possibility highly improbable was the fact that a testes-carrying female virgin would have a very hard time getting pregnant in the first place: because people with testicular feminization cannot make eggs and have no womb in which a placenta could form, they are sterile.

In theory, Mary of Nazareth might have been a genetic

chimaera, rather like Jane of Boston – except that, to have any chance of fertilizing herself, Mary would need to have been formed from a set of twin embryos, one male and one female, who fused into a single body while maintaining both sets of chromosomes – Y and all. This is an intriguing thought experiment, in Berry's view, but it is also an unlikely scenario, not least because Mary's body would have to achieve a truly miraculous balance between male and female hormones and reproductive organs.

It's important to note, however, that in a purely physical sense this kind of sexual ambiguity is actually not that uncommon. As many as one in a hundred people are born with bodies that differ from the standard male or female package, and one in 1666 people are born with sex chromosomes that are not XX (normal female) or XY (normal male); one in a thousand carry XX chromosomes as well as a Y. Far more rare, and for our purposes, more interesting, are the one in every eighty-three thousand people who are born with ovotestes – gonads that are part ovary and part testes.

Indeed, another suggestion is that having ovotestes could explain how Jesus could have been born to a Virgin Mary. Genetically, Mary would carry two X chromosomes and appear to be a normal female – superficially. She would have had breasts (though she may not have been able to lactate), a uterus, Fallopian tubes, and a vagina. Her clitoris, though, would be enlarged, approximating a small penis – reminiscent of Aristotle's hyenas. This larger clitoris would have been due to her anomalously high testosterone levels – produced by the incomplete testes inside of her perfect female form.

Ovotestes result when a woman inherits from her father an X chromosome carrying *SRY* material, which is normally located on the Y chromosome. When cells divide – the process by which any fertilized egg multiplies into the millions of cells that make up a living animal – the cells' chromosomes copy themselves in order to populate the new cells with genetic information. In this replication process, mistakes can and do happen; genes are exchanged between chromosomes quite regularly, and sometimes they are cut and pasted on to chromosomes where they are not supposed to be. So if a person has two X chromosomes, but one of them has acquired essential male genes from a Y chromosome, this would make her completely male.

For the occasion of a Virgin Mary, we would have to go a step further, and imagine that when Mary was a developing embryo, her abnormal X chromosome in most of her tissues lost its male genes. Most, but not all. If the X chromosomes carrying *SRY*-containing genes were retained in some of those cells destined to become gonads, this would make Mary a genetic and sexual mosaic; she may even have had a beard. And depending on the balance of her mishmash of hormones, it would 'simply' be necessary for her ovotestes to produce both sperm and eggs when she reached puberty, and for these simultaneously to travel down the Fallopian tubes, and for the sperm to fertilize the egg, and for the fertilized egg to implant in the uterus. Voilà: virgin birth.

This may sound far-fetched, but just such a case of hermaphroditism was described in 2000. The child, a one-year-old girl from Mexico, was reported to have ovotestes, and to have all the ingredients that might begin the chain of events necessary for an eventual virgin birth. But given her young age at the time of examination, it was not clear whether she would be able to produce viable sperm and eggs at puberty – a question that might very soon be answered, given her date of birth.

⦾⦾⦾

You would be forgiven for thinking that these scientific scenarios are no more plausible than a miracle would be. Indeed, (rather like Monica Jones) misunderstanding, misdemeanour, or even mistranslation may be the real explanation behind the baby Jesus's birth. The great irony, as we have seen in the previous chapter, is that it may, in fact, be almost more sensible to reproduce without males, because of all the drawbacks of sex for females.

Nature has also shown on several occasions that mistakes in the DNA have allowed many animals that normally reproduce through sex to do without it. Sex is, after all, something that evolved only once, which means that those animals today that have the capacity to reproduce without sex regained this capacity relatively recently, through mutation. Switching between sexual reproduction and self-reproduction should, therefore, bring with it some evolutionary advantages. For example, imagine that a mutant animal is born, quite by chance, which is able to reproduce all on its own. This asexual female could give twice as many genes to its offspring as compared to a female having babies only through sex, providing a better chance for these genes to survive. The asexual animal's descendants would get to keep those genes that made their mother (or grandmother) well adapted to the environment – there would be no random mixing and matching with a father's DNA. If that mutant female was particularly well suited to the environment, its descendants would quickly take over. And in fact, this has happened, creating species in which there are no longer any males at all.

The classic example is the whiptail lizard. In one species of whiptail, known as *Cnemidophorus uniparens*, there are no males – and there never have been. The lizards were created when a male and female of two different whiptail species mated, forming a mutant offspring that could reproduce without sex; the eggs of the animal develop into female baby lizards via parthenogenesis. Curiously, to activate fertilization, the females still need to 'mate', and to look at, it seems just like the elaborate ritual one witnesses when a male whiptail courts a female. First, the 'courting' female lunges at and bites her targeted female. The attacked female is at first defensive, and attempts to bite back, but this behaviour rapidly dissolves into a more passive stance. The aggressor then grips the responsive female's tail or leg with its jaws, and then mounts it for two to three minutes. While the lizard is on top of its target, it will intermittently rub its cloaca (that's the one orifice that serves as the sole opening for the lizard's anus, genitals, and urinary tracts) against the back of the passive female, stroking the back and neck with its jaws and forelimbs. The active female then grasps the back of the neck or shoulder of the passive female in its jaws and begins to curve and force its tail beneath the other's tail, so that the cloacae of both are brought into close contact, somewhat as a male lizard would do in order to erect one of its two penises through its own cloaca. Once the orifices of both females are in contact, the courting female shifts its jaw to grip the lower half of the mounted female's body. This forces the couple to adopt a contorted posture characteristic of mating lizards of opposite sexes. Now, if you were to dissect lizards post-'coitus', you would find something quite interesting: the female doing the courting has ovaries containing only small, or immature, eggs, while the passive female hosts several large eggs, each ready to develop into an embryo – almost as though the eggs had just been 'fertilized' by mating with the other lizard. It's not known how the courting activates the egg to divide.

A number of other all-female species have been created by chance, when distinct but related male and female ancestors happened to mate. These include species of snails, crustaceans, and insects, including weevils, stick insects, and grasshoppers (not just aphids). Among other animals, the ability to have virgin births has occurred spontaneously, as happens quite naturally, for instance, in around one in ten fruit-fly eggs.

But for animals for which there are no males in the species, reproducing can be a little tricky at times. One tactic that has developed to get around this is known as kleptogenesis – where members of some all-female species, such as mole salamanders (*Ambystoma*), 'borrow' sperm from the males of related species in order to stimulate egg development, but without allowing the sperm to contribute to the genetic make-up of the offspring. The borrowed sperm kick-starts the eggs into action, and even fuses with the egg nucleus; however, unlike when a human egg is fertilized, the father's and mother's genomes do not combine. Occasionally, the father salamander's genetic attributes will show up in some offspring, but only the mother's DNA is transmitted to the next generation: when the offspring produce their own eggs, the chromosomes they got from the father are dumped and replaced by material stolen from the next sperm donor.

It is among insects, though, that virgin births probably exhibit the most diverse array of genetic strategies of animal reproduction, including some extremely rare mechanisms. Take the electric ant, *Wasmannia auropunctata*, a tiny ant that lives deep in the rainforests of Brazil and French Guiana, though it has now spread around the globe. Commonly known as the 'little fire ant', it is ranked among the world's most invasive species. Some populations of the ant reproduce through normal sex. But for others, the females can make babies without the males – through a rather unusual process. The virgin females produce males, which then fertilize the eggs, though biologists do not understand how they do so. In any case, the fertilized eggs grow into more females – workers that are all sterile.

What is especially odd about the little fire ant is that it isn't just the females that clone themselves; the males can do it too. This indicates that virgin birth in little fire ants probably developed spontaneously, as a result of mutations in their DNA. The males of the species are able to clone themselves by eliminating their maternal genes, so that they only ever pass on their father's genes – the reverse of the scenario when females produce offspring through virgin birth. When scientists first discovered this detail, they assumed that the female DNA was discarded from the egg during fertilization, just as happens in reverse with sperm in female virgin births. It seemed as though the ants were waging a war of eugenics – the females dumping male DNA, with the males doing it right back.

In further studies, it has been revealed that things are not so simple (or so very human). The male little fire ants appear to be clones, without any of their mother's DNA, but this trait appears to be controlled by the females – or more specifically, by the queen ant. Only males mated to the queen can become clones of their fathers, which is to say that, if a son ditches its maternal DNA, it is because the mother has dictated it to do so.

For the little fire ant, it seems there would be no male clones if the females themselves were not able to have virgin births. When the cells in a queen divide to produce parthenogenetic eggs, there is not an equal division. One of the two 'daughter' cells takes the DNA-containing nucleus in its entirety, while the other takes just the cell's cytoplasm. So a queen lays some eggs that have a complete set of DNA and require no fertilization (and which become sterile daughters) and lays eggs that are 'empty', with no nucleus or DNA. Should these empty eggs be fertilized by sperm, the only DNA to be donated to the embryo comes solely from the male – there simply is no genetic information from the mother ant in the egg.

In other species, parthenogenesis is triggered by the sort of hazard that some scientists believe gave rise to sex in the first place: a contagious infection. Believe it or not, this means that you might be able to 'cure' some animals from the plague of virgin births – animals like stingless wasps (*Trichogramma*). These tiny parasites make their living with some rather cunning tricks. The females attack the eggs of moths and other species, injecting their own eggs into those of their unsuspecting victims. In some cases, they find newly hatched eggs via some quite sophisticated chemical espionage. The wasps can sense anti-aphrodisiacs – the pheromones that many male insects pass to females to signal that mating has taken place, and which makes these females less attractive to other males, in an attempt to keep other males' DNA from competing in a race to fertilize the egg. This may be a useful tactic to employ when there is competition for females, but it's also a communication system prone to sabotage by the stingless wasp. By following the scent of these pheromones, the wasps can locate a female butterfly, moth, or other insect that has just mated and is poised to lay a clutch of eggs. The female wasp then hitches a ride on her hostess insect, until the host lands on a plant to lay its eggs. The parasitic passenger hops off and quickly injects its own eggs inside the fresh host eggs, so that when the wasp's larvae develop, they are perfectly placed to eat the contents of the victim eggs. And these host insects are not the only victims. For the female wasps themselves play host to another parasite, one yet more brilliant. When this parasite infects its target, the males of its host species become infertile, or are killed while they are still developing as embryos. And as the males are eradicated, their females begin having virgin births. This is probably how stingless wasps began to reproduce in this way – and why they may yet be kept from transforming into an all-female species.

The parasite responsible for infecting the stingless wasp is the *Wolbachia* bacterium. *Wolbachia* live in the eggs of the host wasp, through them passing from mother to young. But these tiny wasps are not their only targets: *Wolbachia* are among the most abundant and remarkably widespread of all parasitic bacteria, and the number of species known to be infected by the bacteria is increasing rapidly. Just how far afield *Wolbachia* are distributed is uncertain, but so far, they have been found in over seventy-five percent of arthropods, which include eighty insect species, seventeen isopods (a category of crustaceans that includes the woodlouse), many spiders, and one type of mite. It is also likely that there is probably an equivalent proportion of infections among nematode worms. And there is a good chance that it will spread.

If this doesn't seem like much to be concerned about, bear in mind that between them, arthropods and nematode worms comprise something in the order of 99.99 percent of animal species in the world. From cockroaches to termites, dragonflies to ladybirds, woodlice to worms, *Wolbachia* wreaks havoc with ovaries, testes, eggs, and sperm in many ways – not all of which are entirely clear. What is clear is that *Wolbachia* need eggs; they live in them. This means that the bacteria must be assured of a good supply of eggs – and in this respect, males are dispensable. So *Wolbachia* has found a way to make the eggs of stingless wasps develop without any sperm involved.

When the wasps reproduce through parthenogenesis, they produce only daughters, of course. But in 1990, scientists found that they could make females and males start having sex again, and that they could induce mothers once again to have sons. All it took was a dose of antibiotics.

Wolbachia may be the most prevalent bête noire of male insects, but it is by no means the only parasite to play this game. There is a parasitic fungus, *Ichthyophonus hoferi*, that causes parthenogenesis epidemics in fish worldwide. Among its targets is the green swordtail (*Xiphophorus helleri*), a species in which the female fish is able to turn into a sperm-producing, fully functional male, indistinguishable from a true male. In the fish, the fungus causes haemorrhages, destroys muscle, and rots fins and skin – so there is reason to worry about its effects. One of the toxins it produces also seems to make eggs develop without sperm. In addition, there is a large and diverse array of microorganisms that alter the ratio of males to females in their host populations, usually by killing or feminizing the males. These include protozoa, which affect mosquitoes and amphipods, an order of tiny crustacean, most usually less than ten millimetres in size; spiroplasms, which affect fruit flies; enterobacteria in wasps; and *rickettsiae* in ladybird beetles. For most of the planet's species, the sexes and reproduction are not always what they seem.

◎⊙◎

Then, there are those few animals that normally would have sex to reproduce, but don't strictly have to do so. In recent years, the list of these animals has been steadily getting longer. It's not that virgin birth as a phenomenon is necessarily increasing in the natural world; rather, that scientists have simply started to realize that some very large animals – not just insects – are able to reproduce without mates, should they need to.

On 14 December 2001, the list was extended in an exciting new way. Before then, virgin births had been documented in all of the jawed vertebrates – that is, animals with backbones and jaws. The exceptions, therefore, were the mammals and cartilaginous fishes, the latter of which include sharks and rays but also the aptly named chimaeras – peculiar-looking deep-sea ratfishes and egg-laying ghost sharks that have glowing eyes and snouts like an elephant. But that December, an adult bonnethead shark gave birth to a normal, live female pup at the Henry Doorly Zoo in Nebraska.

The adult shark had been captured from the Florida Keys when she was not yet a year old. For the bonnethead shark

(*Sphyrna tiburo*), that meant it was two years away from hitting puberty. Once captured, it had been raised in a tank at the zoo, in the company of only two other sharks – both female. Now, sharks of the hammerhead family, of which the bonnethead is a member, normally can store up sperm from a previous mating for as long as five months. But that would not have been possible for this female, as it was so young when it was caught. The other possibility that the zookeepers considered was that it may have both male and female genitals, as some humans do. But there again, it didn't have any claspers, a kind of penis that a shark that is biologically both male and female could use to fertilize itself. There was only one other way a pup could have ended up in a tank full of adult female bonnethead sharks. And genetic tests later confirmed that the pup was in fact a virgin-born animal, the first of its type to be identified.

For obvious reasons, it is extremely difficult to spot a virgin birth in a wild animal population. Unless you are very lucky, and stumble on a female-only group that cannot possibly be accessed by males – the shark in a tank – reproduction by parthenogenesis in an animal that usually has sex is easy to miss. So it's not surprising that the next reported virgin shark mother was once again found in captivity, seven years later, at the Virginia Aquarium, in Virginia Beach. This time the mother was an eight-year-old blacktip shark (*Carcharhinus limbatus*) named Tidbit; unfortunately, her parthenogenetic offspring was only discovered during Tidbit's necropsy. Like the bonnethead shark mother, Tidbit had been caught in the wild when very young, less than a year old, and had reached sexual maturity in a tank at the aquarium. The shark had died during a routine physical examination – nothing had seemed amiss. When the body was cut open, it was found to be carrying a pup that was thirty centimetres (one foot) long and nearly full term. Like the bonnethead baby, DNA analysis revealed that the female blacktip foetus did not have a shred of genetic material from a father. There couldn't have been, as there were no male blacktip sharks in the tank for the entire eight years that Tidbit had lived there.

Sharks are not the only new animals joining the parthenogenetic ranks. Two years before Tidbit's examination, two virgin births in Komodo dragons at zoos in the United Kingdom were announced; since then, a third has occurred in Kansas. Komodo dragons (*Varanus komodoensis*) are the largest of all lizards; the adults can grow to a length of three metres (ten feet) and can weigh more than ninety kilograms (two hundred pounds). Yet, as wild Komodo populations have become smaller and more fragmented, these legendary creatures have come under threat of extinction. Today, there are fewer than four thousand Komodo dragons remaining in the wild, of which fewer than a thousand are believed to be mature females.

Because of these worrisome figures around the species' future, at least fifty zoos had begun participating in an international breeding programme by the time the virgin births were discovered; it could be said that quite a few people were paying quite a bit of attention to Komodo sex life. In all of Europe, there were only two sexually mature female lizards: Flora at Chester Zoo and Sungai at London Zoo. Both had been bred in captivity, and were therefore crucial to the success of the European breeding programme. But what the zoo staff involved had not realized is that female Komodos can reproduce without a male.

It's not that Komodos need males to lay eggs – much like chickens, it has long been known that some of the eggs Komodos lay will be unfertilized. But in 2006, the then zoo-keeper in Chester, Kevin Buley, took a clutch of twenty-five eggs that Flora had laid and incubated them 'on a zoo-keeper's whim'. It was a serendipitous moment for science, and, indeed, for Komodo conservation. Flora had never been with a male dragon. Yet of the twenty-five eggs that Buley fostered, eleven looked just like normally fertilized eggs would. And in January 2007, eight hatched into healthy male Komodo babies. The three eggs that didn't make it had collapsed early during the incubation, but they proved useful in providing embryonic material for genetic fingerprinting. Through such analyses, Flora was proven to be both mother and father to the surviving eight sons. Similarly, Sungai in London Zoo laid twenty-two eggs – two and a half years after her last contact with a male. Nearly eight months later, four of these eggs hatched, producing healthy sons. Sungai subsequently successfully mated with Raja, a male also housed at the zoo. Two months later, Sungai laid a second clutch of six eggs, only one of which hatched. Sungai has since died, but at Chester Zoo, Flora has been set up with Norman, a two-metre (seven-foot) male in whom the female has so far showed no interest whatsoever.

Komodo dragons and sharks aren't like whiptail lizards; they aren't all females, and they don't behave – as has been said of whiptails – 'like lesbians'. Flora the dragon and Tidbit the shark only experienced virgin births because they had no males to mate with. Theirs was an artificial situation, because they were living in zoos, but sometimes the situation in the wild isn't altogether different. Poaching and human encroachment have decimated the population of Komodo dragons in the wild. Sharks are also increasingly overexploited: in the north-west Atlantic, there have been rapid declines in large coastal and oceanic shark populations – over a seventy-five percent decline, in fact, in the past twenty years or so.

Usually, hammerhead sharks in the wild have litters of around fifteen pups, and blacktips have four to six. Of the two sharks that were found to have had virgin births in captivity, both only had one pup. Having babies without a mate means that the offspring won't enjoy as much genetic diversity as they would if they had a father's genes too. And low genetic diversity is almost always a bad thing – something to be avoided in a population struggling for its existence; if every member of a family line has the same DNA, and that DNA is not well suited to the existing environment, it could spell disaster. Then again, a Komodo mother creating a male with which it can breed doesn't appear to be such a bad option, when faced with extinction. And it's not such a bad option to have just one shark daughter, instead of six, if the alternative is not reproducing at all.

Whether in situations where males are sparse, or simply as a matter of course, it is perfectly possible to create life from eggs alone – it's a strategy that females of many species have long exploited. And indeed, parthenogenesis in animals has even been exploited, in a more domestic context, by us humans.

⦾⦾⦾

After the war, the US Department of Agriculture and British animal research units sponsored some interesting experiments with the ostensible aim of improving the efficiency and sustainability of animal husbandry. Then, in 1952, scientists at the Beltsville Agricultural Research Center in Maryland discovered parthenogenesis in turkey eggs. Immediately, selective breeding programmes were launched in an attempt to intensify the trait in certain lines of turkeys and chickens – it was the hunt for the ultimate breeder.

Those turkeys that showed the greatest predisposition for reproducing without sex were crossed, and were a runaway success – the Beltsville Small White breed. The percentage of virgin births in the turkeys increased from nearly seventeen percent to around forty-five percent in the space of a mere decade – a mote of evolutionary time.

Interestingly, parthenogenesis in the eggs of both chickens and turkeys was notable for the seeming lack of cell organization within the early embryos. A chaotic, multi-layered mess of cells would develop, whether it was grown in a hen's oviduct or in an incubator. This made the scientists at Beltsville wonder whether an infection could have triggered the birds' new mode of reproduction – an avian version of *Wolbachia*.

As a culprit, the researchers suspected a particular group of viruses, Rous sarcoma retrovirus, fowlpox virus, and Newcastle disease. Each of the viruses was found to enhance parthenogenesis; in other words, they appeared to stimulate the egg to develop into an embryo, an effect that, as with *Wolbachia*, persisted in the eggs of the daughters and granddaughters produced through virgin birth. Unlike with *Wolbachia*, however, virgin birth in the turkeys was not 'curable'. If an already vaccinated turkey was infected with the fowlpox virus, the incidence of parthenogenesis in the eggs of that same hen increased markedly over the level recorded for her eggs before she was vaccinated – the opposite to what the scientists had expected.

Still, the data indicated that parthenogenesis could be boosted by selective breeding. The team could identify the 'high-incidence' turkeys, those recorded to have the highest predisposition for virgin birth, and keep the line going, without introducing sperm. The finding was backed by studies of fruit flies, among which it was observed that both males *and* females could transmit the parthenogenetic trait to their offspring. Through cross-breeding these flies, scientists increased the rate of parthenogenesis by around thirty-four times, compared to unselected fruit flies, in just twelve generations.

That, of course, leaves the question: if the birds and the bees can do it, why couldn't *we*?

4

THE CONCERT IN THE EGG

The history of man for the nine months preceding his birth would probably be far more interesting, and contain events of greater moment, than all the three score and ten years that follow it.

Samuel T. Coleridge, annotation to *Religio Medici*, 1802

It was March 1984. In a small town in Lesotho, a fifteen-year-old girl walked to a local bar to start work for the day. At some point during her shift, her new boyfriend showed up. Unfortunately, so did a jealous ex-lover – just in time to catch her with her boyfriend's penis in her mouth: to the ex's eye, in flagrante delicto. The details of the fight that ensued are unclear, but it seems the girl tried to protect her boyfriend (and herself) from attack, because she ended up with lacerations on her hand; she also sustained two stab wounds to her stomach.

Just over nine months after the doctors at Mafeteng District Hospital stitched up her wounds, the girl returned. This time she was much larger in girth and complained of acute, but unexplained, pains in her lower abdomen. The possibility that she was pregnant had never occurred to her – not because she was naïve, but because she was well aware, as her doctors were soon to discover, that she did not have a vagina.

In one sense, having sex and getting pregnant is as straightforward an event as any. You need a male and a female, they have sexual intercourse, sperm meets egg, and, some time later, a baby is produced. All of this we take for granted, though not in a glib sense; as we have seen, much can go wrong along the way, and sometimes babies are born with debilitating diseases and sometimes they are miscarried or emerge stillborn. Yet, in all these cases, we assume that at one point a sperm met an egg, thereby beginning a pregnancy. But step back from that assumption and consider the circumstance within which sperm meets egg, and even whether or not sperm meets egg at all.

The Lesotho teenager was treated by Douwe Verkuyl, who reported on her predicament in the *British Journal of Obstetric Gynaecology*. Dr Verkuyl suggested that the pregnancy may have been the first recorded case of 'oral conception': the knife wound to the stomach may have allowed the sperm the girl had swallowed to find its way from her gut to her womb. As unlikely as it may seem, this is a strategy that has seen precedents in nature. There are animals in which fertilization can be achieved artificially by injecting sperm into the abdominal cavity, from where they swim down a Fallopian tube to the egg. In fact, for some, like the blood-sucking African bat bug, impregnation via the abdomen is the standard mode of operation. Even though the females of the species have not one but two vaginas – one real, one fake – the males use neither. Instead, a male will stab a female's abdomen and inseminate the female's blood. Of course, just because a relative of the bed bug can do it doesn't imply it can happen in humans too. But Verkuyl noted two conditions that she thought made the scenario more likely: when her young patient arrived at hospital after the fracas, her stomach was devoid of food (and the gastrointestinal juices that are produced to digest it) and saliva itself tends to be alkaline (it has a high pH). These things, she thought, helped the sperm survive what would have otherwise been the hostile, acidic environment in the stomach.

Shortly after the child's birth by Caesarean section, the families of the girl and her boyfriend exchanged cattle to seal their union. But the girl's son, conceived from that gruesome and bizarre violent encounter was to be her only child. Around the time her son reached the age of two and a half, the girl began to suffer crippling pains, because her menstrual blood, which had no outlet, was collecting in the cavity of her uterus. Unable to stop her periods through drug treatments, doctors ordered a hysterectomy, the removal of the uterus. 'By that time,' Verkuyl wrote, 'the son looked very much like the legal father,' a fact that 'excludes an even more miraculous conception'. By which, one assumes, the doctor was alluding to the conception of a child without the involvement of sperm at all.

Mission impossible? It should be. Yet, some animal eggs regularly manage to achieve virgin births – not just tiny insects, but large vertebrates, including fish, birds, and reptiles. And, in fact, human eggs can work on their own too.

◉◉◉

Most animals have eggs with a lot in common. Of the billion or so cells in our bodies, the egg is the largest cell that animals have. Among most amphibians and fish, an egg is about as big as the full stop at the end of this sentence. If you divided that full stop into one hundred pieces, most other cells in their bodies would be about the size of just one of those bits. Human eggs, by comparison, are the size of ten pieces. Reptiles and birds, of course, have immense eggs – each egg in the cardboard carton you bring home from the supermarket to scramble up for your breakfast is essentially a single egg cell.

The mammalian egg is the site of a number of quite extraordinary biological processes, not least of which is the way the egg itself is produced. The first person to figure out that there was something unusual about eggs was Edouard Van Beneden, in 1883. At the time, Van Beneden was studying how the cells in *Ascaris megalocephala*, a worm parasite of horses, were made. By throwing live female worms into containers filled with diluted alcohol and leaving them there to 'stew' for several months, he was able to dissolve the worms' cells enough to reveal their components. (This method of damaging cells is still used in labs to isolate DNA and RNA; it's also the reason why alcohol wipes are effective at killing bacteria.) Van Beneden observed that the worms had four chromosomes in almost all of their cells. In their eggs, however, there were just two chromosomes. To Van Beneden, this made no sense: the eggs were made from cells with four chromosomes, so two chromosomes had seemingly disappeared. He then noticed that the mother's and the father's chromosomes came together in the fertilized egg, thus producing baby worm cells with four chromosomes. The contradiction puzzled him, and he did not publish any further work on the subject.

A year later, Van Beneden's work came to the attention of the German biologist August Weismann. Alas, the esteemed professor was suffering from eye trouble. After many years of conducting his groundbreaking scientific experiments in chemistry, biology, and medicine, he could no longer look down the microscope for himself, and he was forced to turn his attention to theoretical questions. Still, he was not ready to give up experimentation altogether. Weismann asked his janitor to carry out the logistics of his experiments and assigned his students to do the microscopic analysis. His wife, Marie, would read scientific papers aloud to him, so that he could keep up with the latest ideas. Among the papers Marie recited was Van Beneden's work on chromosomes in worm eggs. As Weismann listened to Van Beneden's extensive observations, and Marie's descriptions of the paper's drawings, he came up with a theory: a very special division happened exclusively in the sex – or germ – cells, that is, in eggs and sperm.

For every cell in your body that isn't an egg (for a woman) or a sperm (for a man), making a new cell is a simple case of copying the chromosomes, separating the copies into two lots, and then distributing the original set and its copy equally into two new cells. This process is called *mitosis*. Think of it as making a photocopy of some pages: you separate the original pages from the copied pages, keep the originals for yourself, and give the photostats to a colleague.

Weismann realized that a different division must also occur in order to make a sex cell. Whatever number of chromosomes there were in the original cells, these would need to be halved, resulting in that sex cell with only one set of chromosomes. He had discovered *meiosis*, a process that only ever happens in eggs and sperm, and the thing that makes sex exciting, in evolutionary terms; meiosis ups the ante in the grand gamble of reproduction. The word 'meiosis' is derived from the Greek for 'diminution', because, as Van Beneden and Weismann observed through their microscopes, duplicated DNA from a sex cell that is dividing is diminished by half in the new cells that are produced. (The devices available to them were not quite sophisticated enough to demonstrate that, in reality, meiosis achieves far more than that.)

Whereas most cells of our body contain one maternal chromosome and one paternal chromosome, each copied as precisely as possible, an egg or sperm must contain only one chromosome strand, and the copies of the chromosomes that end up in the egg or sperm are not simple duplicates of the strand in other cells. The process of meiosis physically shuffles and exchanges information between the two chromosome strands. To do this, the double helix of the chromosomes breaks, and the broken ends physically move across each other, swapping genes before the double helix re-forms. This is a much greater challenge than the usual process of cell division, because genes must be matched, sorted, scrambled, redistributed, and realigned. From this, a unique combination of genes is born. It is different to the gene combination found in either parent, different to the one inherited in every other body cell, and peculiar to the offspring that may be created when this sexual cell fuses with a mate's. It is for this reason that no two children born to the same parents, unless they are identical twins developed from a lone fertilized egg, are genetically the same.

At the end of this complicated process, it is not just how the chromosomes are divided up that makes the egg especially unique. Inside the egg, there is also a cellular 'soup', which separates into two grossly unequal parts. The disproportionately smaller of the two parts helps to reduce the number of chromosomes until only one set of the two is left. Ultimately, that smaller part degenerates while the larger one sticks around to become the egg, ready and waiting to be fertilized.

This rudimentary, immature egg will then undergo a second meiosis, just as complicated as the first. Another unequal division of a cellular soup produces a tiny cell and a fully mature egg. Like the last one, this tiny cell is usually destroyed, but not always. In the fruit fly *Drosophila melanogaster*, this tiny cell sometimes 'fertilizes' the larger, mature egg to create virgin-born fly offspring, essentially using a part of the egg to stand in for sperm. In humans (and almost all vertebrate animals), right in the middle of this second meiosis, the egg stops dividing and enters a biological holding period, known as *prophase I*, in which it can remain for an extraordinarily long time. In frogs, this phase can last several years; in humans, several decades.

When a girl enters puberty and starts ovulating, the egg will resume its monthly meiosis. But there is another catch: the egg will be blocked from maturing further or transforming into an embryo until and unless some sperm show up. This block on development is dramatically named *metaphase II arrest*. Eggs need to be activated to start their dividing, and activation usually happens with fertilization – the fusion of sperm and egg. At least, that is, when things are proceeding normally.

◎◎◎

Moment by moment in the course of your life, cells in your body are dying off. Before they do so, they divide and give rise to 'replacement' cells just like themselves – a skin cell divides into two new skin cells; a liver cell into two liver cells – which is how the body doesn't dissolve into non-existence. But when eggs divide, they can give rise to every cell type that exists in the adult, creating, over a series of cell divisions, a complete new individual – sometimes, in a matter of days. No other cell can match this feat.

A beautifully orchestrated concert ensues in an animal egg after fertilization occurs. Like a pool bursting with the elegant and energetic motions of a team of synchronized swimmers, molecules interact and cells cluster and move around to the very spot where they will be called upon to shape a new creature in early, miniature formation. Soon (in humans, about fifteen days later), the early embryo organizes itself from a simple ball of cells into an organism made up into what are called *germ layers*: the *ectoderm* (the 'outer skin' in Greek), the *mesoderm* (the 'middle skin'), and the *endoderm* (the 'inner skin') in all vertebrates. These skins are literally the layers that build us, and are responsible for forming all the structures and organs present in a fully developed animal body. It is now that a recognizable body plan begins to be laid out.

The endoderm, the innermost of the three layers, forms a simple tube, which will eventually become the digestive tract, connecting the mouth to the anus. The tube will differentiate into parts as diverse as the pharynx, which helps us to speak; the oesophagus, the 'entrance for eating'; the trachea, or windpipe; the salivary glands; the liver; the pancreas and certain glands of the pancreatic system; and even the lungs – the respiratory and digestive systems being intricately connected. The mesoderm gives rise to the muscular and fibrous tissues – the muscles, including the heart; connective tissues, cartilage, bones, bone marrow, blood, and the epithelia that line the blood vessels; the lymphatic vessels and lymphoid tissues; the reproductive organs and the urinary system; and the notochord, a column of tissue that bisects the embryo into left and right sides, and which later develops into the vertebral column. The ectoderm becomes the brain and spinal cord, via a process in which a part of the layer rolls up into a tube and pinches itself off from the rest. As it pinches off, some ectodermal cells escape into the mesoderm, where they form part of the nervous system as well as the pigment cells of the skin. The rest of the ectoderm envelops the embryo with what will become the epidermis – the outer layer of our skin – complete with sweat glands, hair, nails, and teeth.

An egg trying to make all this stuff on its own is up against a number of natural obstacles. For one, an egg has only one set of DNA, but its offspring requires at least two, to get that right number of chromosomes. Second, to start the process of separating, copying, and dividing up its chromosomes, the egg needs some centrioles – barrel-shaped cellular structures, provided by sperm, that help to move the chromosomes around during cell division. Third, at some point along the way to becoming an embryo, the egg will face the roadblock of metaphase II arrest.

And for mammalian and marsupial eggs, there is a fourth challenge: evolution has locked some genes so they just won't work for creating offspring. Still, some human eggs have gone solo – or perhaps it's more precise to say that they have gone rogue.

The main evidence for the human egg's capacity to develop on its own comes from *teratomas* – shocking, grotesque cell masses that appear to be an amalgamation of unfinished or discarded body parts. Mature teratomas are a rare form of benign tumour made up of varying combinations of ectoderm, mesoderm, and endoderm tissues. They have been documented in guinea pigs, dogs, cats, horses, marmosets, rhesus monkeys, baboons, and humans. Some teratomas are smooth, shiny balls of skin; others, a bloody fur ball of hair; yet others a lump of raw flesh spiked with perfectly formed teeth. Often, under their skins, they also contain organ systems and major body parts. It may not come as a surprise, then, that *teratoma* comes from the Greek for 'monstrous tumour'.

Ovarian teratomas, which grow from egg cells, have been identified in girls as young as two and women as old as eighty-eight, but they mostly tend to develop in women in their twenties or thirties, or 'late' reproductive age. Studies of twins indicate that the propensity to develop ovarian teratomas may be inherited. These teratomas are quite distinct from the more highly developed growths known as *fetus-in-fetu* – malformed, parasitic twins that grow inside a living person's body. Fetus-in-fetu are the product of normal conception, while ovarian teratomas come from eggs that have never had a whiff of a sperm cell. Essentially, they are unfertilized eggs that didn't or couldn't respond to the signals to stop and restart developing – the usual holding periods involved in readying an egg for reproduction.

The vast majority of ovarian teratomas recorded in humans have gone so far as to develop such features as skin, hair, and teeth. In twenty-four known cases, ovarian teratomas have contained a homunculus – a mini-human, or partial, foetus-like structure, something straight out of Paracelsus. The Latin term *homunculus* roughly translates as 'a structure resembling a miniature human body' and today is used by doctors to describe a growth of tissue that has the features of a human being but which was not produced by pregnancy.

In 2002, a twenty-three-year-old woman was admitted to the Korean University hospital with a huge lump in her pelvic area. It was soft to the touch, and it moved around when prodded. The patient had never been pregnant and had regular periods; her womb and her Fallopian tubes were normal. Doctors performed an ultrasound and found that, in fact, the woman had two lumps, one in each ovary. The masses were removed and dissected. Both had smooth, glistening surfaces and measured about fifteen centimetres in diameter. One of the lumps contained another, smaller cyst, filled with hair intermixed with a greasy yellow substance and some fluid. The other lump was more surprising. Encased in a tortuous network of blood vessels, the doctors uncovered another, smaller growth. There were no muscles, ligaments, or organs inside the homunculus, but from the outside it looked eerily like a tiny, dismembered baby, lying face down, with only a hirsute head and part of its right arm formed. The head was partially split open, and from it spilled a herniated brain. X-rays revealed an imperfect but impressively well-crafted skull, shaped somewhat like a cross between a Spartan and a Samurai helmet. The skull bones included easily recognizable structures, including a cranium and a jawbone.

The following year, in 2003, Japanese doctors operating on a twenty-five-year-old virgin identified the most fully formed teratoma found to date. Once again the outer layer of the tumour was filled with a mixture of hair and fat. Cutting through this mess of cells, the woman's doctors found a solid, hard lump. When the lump was cleaned up, the doctors could see that it was a small, 'doll-like' body, mostly complete. Like any normal foetus, the body was covered with fine, downy hair, but the homunculus was unmistakably deformed. It appeared to have spina bifida, a condition in which the ectoderm doesn't quite finish rolling up into the spinal column (the name is Latin for 'split spine'). Its head exhibited malformations normally seen in babies with holoprosencephalia, which occurs when the forebrain of the embryo fails to divide fully into two normal hemispheres. In the centre of the forehead was a single soft, spherical, fluid-filled 'eye' cloaked by thick, long eyelashes – a disorder named cyclopia, for the one-eyed Cyclops of Greek mythology. This strange foetus had one ear, all its limbs, a brain, a spinal nerve, intestines, bones, and blood vessels – even a jaw, already ruptured by several teeth, emerging from beneath the skin. Paradoxically, it also had what looked like a phallus, positioned neatly between its legs.

◉◉◉

A complete human is built from the instructions spelled out on our forty-six chromosomes. Twenty-three of these we inherit from our mother's egg, and twenty-three from our father's sperm. The egg and the sperm, unlike every other cell in the human body (except red blood cells, which contain no chromosomal DNA), therefore each have only twenty-three chromosomes. When egg and sperm fuse during fertilization, these chromosomes are paired – say, chromosome 15 from your mother's egg will be matched with chromosome 15 from your father's sperm – and form a full double-helix set of forty-six chromosomes in the resulting cells.

If you were then to compare the genes on these two sets of chromosomes, you would find that they are either *homozygous* (encoding the same instructions) or *heterozygous* (encoding different instructions) for certain genes that lead to a child's inheritance. Take *EYCL3*, one of the genes that spell out the colour of your eyes. *EYCL3* is located on chromosome 15 and codes for either a blue or brown tint in the iris. The chromosomes 15 that you inherited from your mother and your father may both carry the blue variant of the gene, in which case you are homozygous for this gene and are likely to have been born with blue eyes. On the other hand, the chromosome 15 you inherited from your father may encode brown eye colour, and the one from your mother may encode blue eye colour, and in this case, you are heterozygous for the *EYCL3* gene and will probably have brown eyes. (The inheritance of eye colour is quite a bit more complex than that, involving several genes and their interactions.)

By this logic, the DNA of ovarian teratomas, coming only from an egg, should be homozygous – it all comes from the mother, after all. But some genes in mature ovarian teratomas have been found to be heterozygous. And teratomas almost always contain forty-six chromosomes, with any outliers involving missing or extra chromosomes – a teratoma with forty-five, forty-seven or forty-eight chromosomes, not the twenty-three available in the egg. The missing or extra chromosomes do affect the development of the teratoma: having three copies of chromosome 13 has also been implicated in the fused brain and 'cyclops' eye features that appeared on the Japanese homunculus. But the teratomas seem to gain or lose chromosomes fairly randomly; some have lost chromosome 13; others have gained an extra copy of chromosome 21, which, in a fertilized embryo, sometimes causes Down syndrome. Not even the sex chromosomes are off-limits: though teratomas nearly always have the XX chromosome signature of a female, a few have been found to contain XXX (one extra X), XXXX (two extra Xs) or XO (a missing X). The one thing that seems to be true of all teratomas, however, is that they never have a Y chromosome, which makes sense, since eggs should never carry this genetic material. Even without a Y, these kinds of tumours have been known to grow prostate tissue, even more often than tumours of the male testis do.

The very fact that ovarian teratomas appear raises many intriguing questions. What makes an unfertilized egg start dividing? How does the teratoma end up with two or more sets of chromosomes, when no other source – say, a father – has contributed to the teratoma's creation? How is it that the teratoma can have two different versions of the same gene if it started life as an egg, which would hold a copy of just one version? How does it grow prostate tissue or phallus-like organs when an egg has merely the X female sex chromosome at its disposal? And how does the egg get around the requirement for other, non-genetic components, such as the centriole, which are usually the unrivalled domain of sperm?

◎◎◎

Human eggs are formed early in life, when a woman is but an embryo, between three and eight months into development. After that, all the eggs can do is wait.

The waiting usually lasts many years – just over a decade or so, until sex hormones begin to exert their powerful effects. Hormones are chemicals, circulating in the blood stream, that act on different target glands around the body. In women, the production of the sex hormones by the ovaries is stimulated by signals from the pituitary gland, a pea-sized structure at the base of our brains. Once puberty hits, the ovaries produce oestradiol, progesterone, and testosterone in a choreographed manner, with levels of the hormones in the blood shifting day by day in the dance from ovulation to menstruation or fertilization. Though it is a form of oestrogen, oestradiol is not a 'female' sex hormone as such; it is also produced in men as a by-product of testosterone, as is progesterone. Oestradiol does, however, play a very important role in female fertility, triggering, for instance, the growth of a variety of reproductive organs, including the vagina and the placenta. Progesterone, too, plays its part, including, it seems, in helping to keep the mother's immune system from rejecting the embryo during pregnancy. With puberty, eggs begin maturing at the rate of approximately one every month.

By around age fifty, give or take ten years, the majority of the eggs that were present at a woman's birth have been released – either discarded during her monthly menstruation or fertilized. Around this time, the ovaries stop acting on the signals from the pituitary, and oestradiol levels fall significantly – to around the same level as is present in men. Progesterone levels also take a dive. This rapid loss of sex hormones in a woman's blood stream is what causes the hallmark symptoms of menopause: hot flushes and sweating attacks; a rise in the risk of heart disease and stroke; and osteoporosis, the 'thinning' of bones.

There are exceptions to the timing of the onset of menstruation and menopause, these critical moments at which the ovaries change their response to hormonal signals issuing from the brain. Just as ovarian teratomas have been reported to be found in octogenarians, post-menopausal mothers crop up every so often. In 1987, a fifty-five-year-old British woman, Kathleen Campbell, gave birth to a baby boy who was verified to be the product of natural conception – the oldest confirmed mother in the UK. Ten years later, a Welsh pensioner named Elizabeth

Buttle claimed to have toppled that record by becoming pregnant naturally at age sixty. After Buttle sold her story to the *News of the World* tabloid, reportedly for £100,000, it came to light that she may have actually been fifty-four, and that she may have been treated at a fertility clinic – neither of which, for privacy reasons, could be definitively confirmed. Mrs Buttle referred to her son as her 'little miracle', and medical experts tended to agree. According to the *Independent*, doctors opined 'that a natural birth to a woman of fifty-four would be exceptional but to one of sixty it would be miraculous'. The legal wife of the baby's father could not be swayed, however; she told the world that, for her at least, miracle or not, the birth was not a cause for celebration.

Pre-teen pregnancies are even less a cause for celebration. But while women who have gone through their menopause must deliver a 'miraculous' egg, pre-pubescent girls simply need some errant sex hormones to activate their more than plentiful supply, waiting for fertilization. And abnormal hormonal activity, including the early onset of puberty, is not unusual and, in fact, is linked to hypothyroidism (when the thyroid gland does not make enough hormone, often caused by a diet lacking in iodine) and several other medical conditions.

Abnormal hormone levels have been known to trigger menstruation at an age when girls are still babies themselves. Take the case of the youngest mother in medical history, Lina Medina of Antacancha, Peru, who had her first period at the age of eight months. At four years old, she had clearly developed breasts and pubic hair. A little more than a year later, in 1939, when she was five years and seven months old, Lina gave birth to a healthy baby boy; she named him Gerardo after the obstetrician, Dr Géado Lozada, who had cared for her. Some in her native town likened her to the Virgin Mary; others believed her child to be the son of the Incan sun god Inti. But, despite the fact that Lina never revealed the identity of her son's father, and sad though it may be, Gerardo was not considered by her doctors to have been conceived without sin.

Similarly, in 1957, a nine-year-old girl was taken into the University of Arkansas Medical Center in Little Rock. Her mother had noticed that her stomach was getting rather big. Examining the girl, the doctor felt a soft, movable lump, which he was convinced was a tumour. To confirm his suspicion, he performed an X-ray. But what he found was not just a lump; it turned out that the girl's periods had started at age eight, and her breasts had started developing the year before. 'Subsequent talks with the patient reveal this not to be an immaculate conception,' the doctor said. Six days later, still in something of a state of shock, he noted in his records, 'I cannot hear the foetal heart beat, but my "ovarian tumour" has definitely kicked me!'

⊚⊙⊚

These extreme cases still involve not just hormones but fertilization, of course. Ovarian teratomas, however, are truly immaculate 'conceptions', coming from eggs that at no stage have been fertilized. And yet, something happens in the body that overrides metaphase II arrest and sends the egg on the path of development, sometimes building remarkably well-developed organs and features. Their origin is inextricably linked to parthenogenesis, but how exactly is the egg triggered to start dividing without first being fertilized? To this question, certain mutant mice may hold the answer.

The structure of most cells in the body can be grossly divided into two areas: the nucleus and the cytoplasm. The nucleus can be thought of as the control centre of the cell. In the nucleus are the chromosomes, which carry the vast majority of the cell's content of DNA, the all-important genetic instructions for the new being. The cytoplasm is a fluid matrix that surrounds the nucleus and all the other *organelles*, or miniature vital 'organs' in a cell, providing the site for much of the cell's chemical activity and manufacture of protein-building blocks. So you can think of it as something like the factory floor to the nucleus's administrative HQ.

During the creation of the cloned Dolly the sheep in 1998, through to the first cloned rhesus macaque monkey embryos in 2007, it was this nucleus-cytoplasm status quo that scientists considered to be essential. In order to clone the sheep and the monkeys, researchers destroyed the DNA-containing nucleus of an unfertilized egg and replaced it with the nucleus of an adult cell. An electric shock was used to activate the egg, instead of fertilization with sperm, and the resulting embryos – which were genetically identical to the adult cell but had no resemblance to the egg donor – were implanted into the womb of a surrogate mother.

Paradoxically, research conducted in the 1960s had indicated that if you take the nucleus from one egg cell and place it into another, the nucleus that was introduced would adopt the behaviour of the host cell rather than the host cell taking instructions from its new nucleus. The cytoplasm was dictating orders to the chromosomes in the nucleus, 'telling' the egg whether or not to divide and mature – a case of the body controlling the brain, as it were, instead of the other way around.

In 1971, Yoshio Masui and Clement Markert of Yale University set up an experiment to work out what exactly in the cytoplasmic soup was pulling the nucleus's strings. They found two powerful ingredients. The first they called *maturation promoting factor*, or MPF, because it puts a cell on the road to mitosis or meiosis. The second they called *cytostatic factor*, or CSF. It is CSF that prevents an egg from developing into an embryo. CSF stalls meiosis in the egg through a delicate communication system of proteins. One of the proteins certainly involved is Mos, which is made before the early egg embarks upon its first meiosis. If Mos is injected into a normally dividing embryo, all cell division stops. After successful fertilization, Mos is destroyed in the cytoplasm, which allows cell division to get going. But Mos does not work on its own. Another protein, called Emi2, also helps to stop an egg from becoming an embryo. All of this intricate chemical activity seems to exist for just one reason: to stop virgin births from occurring. Indeed, *c-mos*, the gene that encodes the Mos protein, is a growth-controlling gene that has the ability, if it is mutated or otherwise unregulated, to cause a tumour to form.

This connection between unregulated tumour growth and very regulated egg growth was tantalizing to scientists. So, in 1994, a team of researchers based at the University of Cambridge and New Zealand's Ruakura Agricultural Centre created mice with a shorter than normal version of the *c-mos* gene. The smaller Mos protein produced from this mutant gene did not work and was unable to order around the cell in its usual way. In many of the mice with the defective Mos, eggs spontaneously divided – parthenogenesis. And one in three of these mutant mice developed ovarian teratomas.

This seemed to be unmistakable evidence that the development of ovarian teratomas is related to mistakes in the *c-mos* gene. Except we know that the Mos protein does not play any significant role in the development of ovarian teratomas in humans. Human ovarian teratomas may come about because of mutations in any of several genes that, in their normal forms, make proteins that hold a cell's development at bay, including Emi2.

What else could make an unfertilized egg start dividing? It has long been known that fertilization by sperm triggers a surge of calcium into the egg. Indeed, in the lab, adding calcium to an egg is routinely used to start parthenogenesis. This offers scientists another candidate protein: calcineurin. Calcineurin is involved in immune system function, putting T cells into action, and mutant mice that cannot produce it exhibit behaviours similar to symptoms of schizophrenia. Calcineurin should be dependent on the presence of calcium to work, but a genetic mutation might allow it to work on its own.

Finally, teratomas must get around the requirement for other, non-genetic components, such as the centriole, which are normally inherited from the father. In many species, including worms, snails, fish, and amphibians, the requirement for centrioles is the main preventative measure against virgin birth. In mammals, however, the process of moving the chromosomes around in the cell is a little more complicated. For instance, unfertilized mice eggs have centrioles, which organize the chromosomes inside. Human eggs also have centrioles, but they do not work, which is why human embryos inherit centrioles from the father. Maybe, once in a blue moon, those maternal centrioles have some say in what's going on in the egg.

◎◎◎

In the Palais des Beaux-Arts in Lille, France, hangs a desolate but fantastical painting, *The Concert in the Egg* based on a drawing by the Dutch master Hieronymous Bosch. In it, you are confronted with an impossibly large egg, flanked by two withered trees – on one hangs a languid serpent, on the other, a wrinkled apple – as if the Garden of Eden had fallen into decrepitude. The egg is cracking open from the many characters it contains: bishops, nuns, simpletons, aristocrats, paupers, the elderly, the ill and infirm, even a monkey playing a pipe. Several of the characters hold a musical instrument: a flute, which was a common phallic symbol; the harp, representing the female sex organs; and the lute, associated with seduction. Fish, birds, monsters, and demons lurk around the egg, but the players seem oblivious. They continue their concert in complete absorption.

In the sixteenth century, when the scene was painted, science had not even realized that there was such a thing as a human egg. And if there was the idea that women might, like hens, have eggs of their own, then it was not much more than wild speculation. But speculation set the wheels in motion, and by 1651, William Harvey, 'Physician Extraordinary' to King James I, was moving away from the realms of folklore and casual observation to form a medicine based on experimentation and precise measurement. Though best known for his seminal research into the workings of the circulatory system, Harvey also devoted considerable time to investigating reproduction, tinkering with chick embryos and, in perhaps his most audacious move, the royal herd of deer.

Against the historical tide of spermists, Harvey claimed himself to be an ovist. Harvey believed that all life came from eggs: not just for birds, which was obvious, but for mammals too. He summed this up in his last work, *Experiments Concerning the Generation of Animals*. It was at odds with the generally accepted Aristotelian view that males contributed the lion's share to the creation of new life through their sperm. After many years of research on the eggs of birds and deer, Harvey begged to differ. He was not able to provide a sound explanation for his ideas about reproduction, as he had done for the circulation of blood. Compared to the circulatory system, mammalian eggs were tiny and posed no small challenge to a scientist using seventeenth-century experimental tools. Only very cautiously, and after great persuasion, did Harvey publish his revolutionary book on sexual generation.

The book begins with a frontispiece, reminiscent of *The Concert in the Egg*, in which Jove sits on a plinth while balancing an egg as large as an ostrich's in his hands. The egg has split in two, and from it escapes an insect, a spider, a deer, a snake, a bird, a lizard, and various other creatures. Leaping out of the egg among them all is a cherubic human baby. Across the egg's shell is scrawled Harvey's hypothesis, *ex ovo omnia* – 'out of the egg, all things'.

Perhaps everything *is* contained in the egg, just waiting to be sprung into life. But although we now appreciate many of the egg's complexities, there is a surprising amount that we still do not understand. What we do understand is that if the beautiful orchestration of genes, proteins, and hormones is disrupted, the egg can give rise to chaos. But although a rogue gene may be enough to kick an egg into forming an embryo, for humans, this is not enough to create a healthy child. However many human features there may be, the tertatomas born of eggs alone are still grotesque caricatures, with no hope of breathing life. There is a switch encoded in the genes of mammals that means that the healthy development of a bona fide virgin birth can happen only in truly exceptional circumstances.

Quite on their own, our eggs can give rise to monsters and to mutants. But mostly what we observe are the success stories – the eggs that are fertilized and grow enough to be born into the world.

5

SECRETS OF THE WOMB

A peace is of the nature of a conquest;
For then both parties nobly are subdued,
And neither party loser.
William Shakespeare, *Henry IV, Part 2*, c. 1600

In the early 1980s, a group of scientists were finding it very hard to convince themselves of something. They knew that a person needs two sets of chromosomes to come together in order to create the amount of DNA that a normal human, or for that matter, any mammal, has. So why couldn't they make a healthy mouse in the laboratory that had two mothers or two fathers – the two necessary sets of chromosomes, though not from the usual two suspects? They had taken some early-stage embryos, removed the DNA that had come from the father, and replaced it with an equivalent set of maternal DNA from another egg. They also tried replacing the DNA from the mother with another set taken from sperm. None of these embryos survived.

We humans could manage to concoct teratomas naturally with their monstrous lumps of skin, hair, and teeth, but that seemed to be the limit of what we could achieve without sex. Mammalian teratomas certainly jump developmental hurdles to develop incredibly sophisticated, if deformed, body parts. Why couldn't a woman be more like a turkey? Why couldn't we just clone ourselves? What was really standing in the way of a virgin birth?

The answer may lie with a feature that no teratoma, in a woman or any other animal, has ever been found to grow: a placenta.

◉⊙◎

On the simplest level, the placenta allows oxygen and carbon dioxide to be exchanged between two organisms: mother and foetus. It is also the medium through which vitamins, glucose, fatty acids, and other sources of nutrition are transmitted to the developing embryo. Yet, despite its essential function in reproduction, the placenta does not develop until adulthood; it is also the only organ to be discarded after it has served its purpose, only to be regenerated the next time it is needed.

Perhaps because of this bizarre cycle of creation and destruction, cultures throughout the world have developed practices, rituals, and myths around the placenta, to account both for its importance and its impermanence. Many animals – including some humans – eat it. Indeed, there are numerous recipes online for anyone who wants to savour one, including such delights as roast placenta (with bay leaves, a tomato sauce, and peppers), placenta cocktail (chilled, with vegetable juice), and placenta lasagne, or bolognaise. You can even dehydrate and use it much like a chorizo.

In some cultures, the placenta is buried with great ceremony after the birth of a child. In Hawaii, this tradition was briefly made illegal, until a law came into force in 2006 that guarantees a woman's right to take her placenta home from the hospital so that she can perform the rite. In Malaysia, the placenta is considered a baby's sibling; in Mexico, its friend, *el compañero* – a good description since, for humans, the placenta truly is indispensable. Without it, humans could not give birth to live babies; it supplies all the things that a foetus in the womb cannot get for itself.

The appearance of mammals, as well as snakes, birds, and lizards from a common ancestor back in the Jurassic period – about a hundred and fifty million to two hundred million years ago – was dependent on the evolution of this remarkable organ. From elephants to elephant shrews, and from dolphins to flying lemurs, the overwhelming majority of the 4600-odd species of mammals alive today develop a true placenta, allowing offspring to emerge from a mother's body with well-developed organs after an extended period germinating in the womb.

Some mammals, specifically marsupials such as the kangaroo and koala, have only a pseudo-placenta, which means that they must give birth at a very early stage of gestation. After that point, the embryo – looking something like squirming larva – crawls from the womb and finds its way to the pouch covering the mother's nipples, where it can suckle milk. The joey continues to develop there, technically outside of the mother's body, for many months. The pseudo-placenta in marsupials does not last long, nor is it very sophisticated.

The strange and complex approach to making babies in marsupial mammals reflects the strange and complex evolutionary history of the organ. In fact, the placenta has actually been 'invented' many times, in different families of animals. Fish have a version, in varying forms, and some sharks have a very advanced placenta, but no species have placentas quite like those found among mammals. Mammalian placentas are extremely complex and structurally diverse, with up to six layers that connect the mother with the developing embryo. Genes have arisen to adapt the mammalian placenta to a range of reproductive environments, to cater for situations as diverse as the twelve offspring expected after a mouse's twenty-day gestation, to the lone calf that results from an elephant's two-year pregnancy. How the placenta evolved to meet these needs, and everything in between, remains a mystery.

The last common ancestor of mammals, birds, and reptiles is likely to have had only the chorion, the outermost of a true placenta's four layers, and a very basic one at that. Today, you can see a chorion when you peel a hard-boiled egg; it is the delicate layer that you find sticking to the shell, just inside. In modern birds and most reptiles, this thin, translucent membrane allows gases to be exchanged between the egg and the outside air. Our last common ancestor would also have had a yolk sac, which in modern birds, reptiles, marsupials, and some fish circulates nutrients to the developing embryo; it would also have had a primitive allantois, another sac-like tissue that gives rise to the blood vessels of the mammalian placenta as well as the umbilical cord. In chickens, the chorion and part of the allantois fuse together, forming a layer thick with blood vessels that pulls calcium from the eggshell to help nourish the embryo. In most marsupials, the allantois develops no blood vessels, since the embryo remains in the pseudo-placenta for only four or five weeks; instead it serves to store waste from the foetus's kidneys, a critical job ensuring that the embryo doesn't abort from toxicity. In other mammals, though, the allantois serves this purpose but is wound up in the wiring of the umbilical cord. Somehow it is these basic layers, the chorion and the allantois, that in mammals became the complex placenta.

Early in human development, after sperm meets egg and the DNA from father and mother have fused, the fertilized egg divides into two cells, as we saw in the last chapter. But the story is not as simple as that: the two cells are not equal. One of the cells is larger, and will continually and rapidly outgrow the other, as it divides over and over again. As the larger cell masses into a solid ball, it is surrounded by a flattened ring of tissue, constructed through the labour of the smaller, more slowly dividing cell. In nine months, it is the solid ball of cells that will be pulled screaming from the womb as a newborn child. The outer ring, called the *trophoblast*, will never become a part of the baby. It is the precursor of the placenta.

As the trophoblast grows into the life-support system for the embryo, in humans it will twice invade and destroy the lining of the mother's womb. The first time happens within a week of conception. Fifteen weeks after conception, the second incursion occurs. This time the cells penetrate far deeper into the uterus, boring one-third of the way into the womb's walls and entwining itself there in a structure that resembles a labyrinth, ensuring that nutrients can be leached from the mother in order to feed the foetus. A cycle of creation and destruction appears even here.

In primates, including humans, biologists believe that, over evolutionary time, the trophoblast gradually infiltrated the mother's womb more and more deeply as the size of the foetal brain grows larger and larger. The brain, for instance, needs about sixty percent of the total nutrition supplied to a human foetus, compared with about twenty percent in non-primate mammal embryos. This may be why the human trophoblast invades the womb twice – something not seen in any other mammal.

The human placenta's progressive and deep invasion into the womb poses a considerable challenge to the mother's body. The mother's immune system should protect her from infections and other foreign threats. But when she is pregnant, her immune system is forced to tolerate certain foreign material – the embryo and the placenta, both of which grow from that first cell that is half derived from the father's DNA. From the body's point of view, these growths are parasites, sucking life from the mother's body for their own existence. Detection by the immune system of such foreign tissue would usually lead to organ rejection, and preventing rejection is a necessity when it comes to making babies.

To ensure their survival, the embryo and the placenta cannot simply suppress the mother's immune system, as this would expose her (and her developing foetus) to the risk of infection – even possibly death. Instead, the trophoblast produces a special subset of MHC class I molecules that protect the foetus from natural killer cells in the mother's womb. These particular molecules work only in the vicinity of the placenta – a neat biological trick. Still, these molecules are not all-important. If you were to destroy them, a mother would not immediately reject her foetus or its placenta. How could that be?

There must be at least one other strategy that prevents full-blown immunological warfare between mother and child. As it happens, the genetic compromise does not seem to have been developed specifically to adapt to life with a placenta. Instead, it depends on the fact that there is another, completely different source of DNA in the body. The cells that give rise to the placenta, and which protect the inner layer of cells destined to become the baby from attack by the mother's immune system, are unique: not only do they exclusively come from the father, some of those genes are not even human. They are the DNA of ancient viruses.

◎◎◎

Among egg-laying animals, which do not have placentas, contact between mother and foetus is very limited, of course. It is the shell that shields the embryo from the mother, and the mother from the embryo, and the shell is created entirely by contributions from the mother. The egg also does not stay inside the mother's body for very long after it is fertilized. Incubating a fertilized egg inside the mother's body required an ingenious ploy. In 1997, Luis Villarreal, a molecular biologist at the University of California, Irvine, wrote an article entitled 'On Viruses, Sex, and Motherhood' in which he recounted his theory of a very clever leap in evolution. In this article, published in the *Journal of Virology*, he proposed that viral DNA played an essential role in the evolution of mammalian pregnancy.

Viruses are among the oldest and most successful life forms on the planet, and Villarreal and others believe that the virus in question would have infected a distant ancestor in our primate lineage as far back as twenty-five million to forty million years ago. When you look at the genome of vertebrates, you find thousands of foreign elements that look a lot like the genetic information harboured in retroviruses, a form of virus that creates DNA out of RNA (opposite to most viruses, which make RNA out of DNA) and integrates this new DNA into its host. Indeed, nearly ten percent of human DNA today appears to be made up of old retroviruses. The most well-known retrovirus is HIV, the cause of AIDS, which shuts down the immune system, but other retroviruses have been linked to tumour cell growth.

Deploying some of the same tactics that viruses use to evade our immune system, the viral DNA in mammals allowed another invader into the body: the foetus. These genes, known as *syncytin* genes, allow the body to protect, nourish, and incubate the foetus, giving it time to mature without the threat of rejection by the 'host' immune system. Without viral DNA, humans and many other mammals might still be laying eggs. And *syncytin* genes are targeted on making the placenta, specifically at the level of the trophoblast ring of cells – exactly where exchanges between the foetus and the mother take place. Importantly, *syncytin* genes instruct cells to fuse with each other. That is, they are able to force cells from the lining of the mother's womb, comprised solely of the mother's DNA, to fuse with cells from the trophoblast, which is designed by DNA from the father alone.

It is important to note that *syncytin* genes work as diplomats rather than combatants in the war between different DNA: they do not affect the embryo's development or the actions of the immune cells; instead, they temporarily cloak the embryo, keeping it from being recognized and destroyed by the mother's immune system. When a particular *syncytin* gene, *syncytin-A*, is disabled in mice, the entire architecture of the trophoblast changes dramatically. Embryos begin to grow, but at a slow rate, and fewer blood vessels form to feed them. The pregnancies invariably end in miscarriage. A faulty placenta does not make for a healthy pregnancy, and this is exactly what the scientists attempting to create fatherless mice were fighting. Their early experiments kept showing that when they tried to produce offspring with DNA originating only from sperm, the embryo struggled to develop; when they did the same with DNA only from eggs, the embryos developed normally but the placenta and other supporting tissues failed to thrive.

But this raises a more perplexing question. Why would a father's genetic contribution be necessary in making a placenta, when viral DNA appears in the genome of both sexes? Why didn't evolution give females the capacity to make a placenta all on their own?

◎◎◎

The evolution of the placenta must have been something of a double-edged sword for our ancestors. While being able to gestate inside an adult afforded unprecedented protection for vulnerable young, mammalian embryos functioned like a parasite on the mother. Apart from the challenges to the mother's immune system, the embryo drained nutrients via the newly designed placenta. This nutrient flow has to be regulated by the body, so that neither mother nor embryo is starved. Ancient viral DNA cannot handle this; new genes with new instructions had to tackle the task.

Not all genes in our cells work all of the time or in all parts of our bodies. Some, for instance, only work in the limbs when a foetus is developing in the womb; some only in the brain of an adult. As this indicates, genes have to be turned 'on' to have an effect, a phenomenon known as *gene expression*. Gene expression can be understood as the process by which the letters of the DNA code are 'read' and start the production of certain proteins, which tell cells (and thus everything in the organism) what to do and when to do it. For some parts of the genome in animals, the expression of a particular gene is determined by whether it was inherited from the mother or the father.

As the placenta gradually evolved in mammals, evolution had to find a way to tell the viral genes and the newer genes when to start working and when to stop. Sometime around a hundred and forty-eight million years ago, certain genes vital for the healthy development of the placenta started to become locked and unusable – coded so that they could never be read, or expressed, since they sometimes mucked up the works. So even though the mother's genome still contains all the genes it takes to grow a complete baby from one of her eggs, only some of them are allowed to function. The same is true for some of the father's genes. This sexual selection in whether a certain gene can be expressed is called *genomic imprinting*.

There is nothing inherently 'wrong' in the coding of these genes that don't work. Imprinting doesn't involve a mutation or a mistake that stops the gene from working – think of it as a padlock that means the gene's DNA cannot be accessed. But just as a door can be opened if you find the right key, imprinted genes can be unlocked, even erased, by different conditions. The process is by no means static. And of the twenty-three thousand human genes that can be expressed by making proteins, only about eighty are ever silenced by imprinting. What is interesting is that many of these genes that are imprinted dictate not what we will look like, but are able to manipulate the growth and nutrition of the foetus in the womb. It seems that when evolution invented sex, it used imprinting as a way of ensuring that the female needed the male to reproduce. The health and survival of any offspring depends heavily on the father's genes for making the placenta, since the mother's genes have been locked. It seems that sperm do more than just deliver packets of DNA into eggs – they regulate pregnancy itself.

Imprinted genes, like viral DNA, are a frontline in a battle: two beings fighting over scarce resources, with some genes trying to ensure the best result for the child at the expense of the mother, and others, for the mother at the expense of the child. The majority of genes that are locked in the mother's DNA but not in the father's directly influence how many nutrients a foetus is able to extract from the mother's body. A father's genes benefit if his offspring are larger and stronger when they are born, because that gives them a better chance of surviving to adulthood and the father's genes being passed on further – for the father, there's no personal risk involved. In contrast, many of the mother's genes that do work at this stage are trying to curb the foetus's growth – to keep those nutrients for the mother. Consider, too, that if every time she became pregnant, the mother could restrict foetal growth, she would secure a better chance of producing more children from limited resources, and she would be less likely to die from complications of childbirth. Evolutionarily speaking, this is to the advantage of her genes, which would have more opportunities to be passed on to a future generation.

This strategy is custom-made for polygamous reproduction. When each female regularly bears offspring of several different males, the mother has an equal genetic stake in each embryo and will achieve the best outcome for her genes if resources are allocated equally to each one; the father is better served, however, if his particular embryos grow faster and extract a greater share of resources from the mother than do the siblings in which he has no genetic stake. So silencing certain genes in the placenta ensures that every foetus has an equal chance of survival. The ability of a father's genes to influence how an embryo acquires resources from its mother is rare, but it does also appear in some plants. In these plants, including maize (*Zea mays*, or corn), the mother nourishes the growing embryo for an extensive period after fertilization, whereas the father experiences negligible costs – just its seed.

In theory, imprinting does not make sense for a monogamous species. A father who intends to have multiple children with just one female partner should co-operate with her for resources rather than try to extract everything he can for the benefit of his genes. Take, for example, what happens when a strait-laced oldfield mouse (*Peromyscus polionotus*) is crossed with its promiscuous relative, the deer mouse (*Peromyscus maniculatus*). To be precise, the oldfield mouse is not strictly monogamous in the wild; it's just that the females don't change partners nearly as often as their polygamous relatives. So when biologists decided to poke into the question of whether any monogamous animals have imprinted genes, the short answer was that they did, in part because they are somewhat promiscuous and had fully polygamous ancestors. Nevertheless, the experiments still yielded some extremely interesting results.

In oldfield mice, the male and the female are about the same size, which is generally the case with monogamous species, and even though polygamous animals usually exhibit a substantial difference in size and appearance between the sexes, deer mice are roughly the same size as oldfield mice. The animals seemed well suited to be mates. Despite this, when a female deer mouse was crossed with a male oldfield mouse, their offspring grew up to be forty percent smaller than either parent. And when a male deer mouse was crossed with a female oldfield mouse, the babies were oversized, bearing enlarged tongues that made it difficult for them to eat and swallow; for the most part, they did not survive. And it was not just that the embryos were overgrown – the placentas that nurtured them were overgrown, too – around six times bigger than in a pregnancy involving two monogamous or two polygamous mice. As a consequence, oldfield mice mothers often died in labour, while trying to push the babies out through the birth canal.

Though both mouse species had imprinted genes, the polygamous females were better equipped to do battle against the monogamous males' genes. The embryos were restricted in taking resources from the mother's body. Similarly, the polygamous males were better able to extract nutrients from the mother for the offspring, building a supersized placenta to increase the foetuses' (and the genes') access to the resources. If there is a mismatch in the genes that are silenced between the mother and the father, however, fatal mistakes can result.

◎◎◎

A pregnant woman's body is constantly negotiating with the foetus over the share of nutrients each one gets. Among the body's main energy-supplying fuels is the sugar glucose. To control glucose, you need to control the hormone essential in the body's proper use of sugar: insulin. And when the body is not producing enough insulin, or becomes resistant to its effects, you suffer from diabetes. Up to fourteen percent of women suffer with diabetes during pregnancy, and although the condition usually disappears after the baby is born, nearly one in five of these women go on to develop Type 2 diabetes within nine years; they may also be at greater risk of developing heart disease. The reason for this lies with imprinted genes.

During pregnancy, the placenta pumps out various hormones that block the usual action of insulin so that the foetus will gain greater access to the glucose circulating in the mother's blood. Effectively, the mother is left unable to control or use her own glucose, making her insulin-resistant, and glucose does not enter her own cells as it should. Glucose levels rise in her bloodstream, and, in something of a vicious cycle, her body needs to produce more insulin to overcome this spike. If it does not, she develops diabetes for the duration of the pregnancy. In adults, both the mother's and the father's copies of the human insulin gene, known as *INS*, work just as well as each other. In the embryo, however, *INS* is one of the small number of genes that are imprinted, so that only the father's copy of the gene functions. The same story plays out for a related gene called *IGF2*, which makes insulin-like growth factor-2. The gene plays a vital role in the growth of the foetus and the placenta: too much insulin-like growth factor-2 makes huge placentas and severely oversized babies – rather like what happens in the pairing of the monogamous oldfield mouse female with a polygamous deer mouse male.

Of course, sugar isn't the only resource over which the mother and the foetus are fighting. By the sixth month of pregnancy, the mother's body has produced an extra 1.4 litres (2.5 pints) of blood to support the foetus's growing needs for oxygen and other nutrients. Pumping this extra blood around the body requires some changes to the woman's circulatory system. Levels of the hormone progesterone increase, to relax and expand the blood vessels in an effort to accommodate the extra flow. In the best-case scenario, this extra rush of blood makes a woman feel unusually hot and involves a drop in blood pressure, which might cause dizziness, or the occasional faint. But when there is a poor exchange of blood between the mother and the foetus, the body has to find ways to push this extra blood in and out of the placenta, and brute force is the answer. Blood pressure rises in compensation. Approximately fifteen million pregnant women experience high blood pressure around the world each year. And one of the reasons pregnant women are constantly having their blood pressure measured is to assess the chances of a medical condition called pre-eclampsia.

Eclampsia in humans was recorded in early Egyptian, Chinese, and Indian medical texts dating as far back as four thousand years ago – not surprising, since the condition involves spectacular, life-threatening complications that would be evident without any knowledge of the interior anatomy and genetic developments involved in pregnancy. If pre-eclampsia isn't spotted and prevented, sudden convulsions can develop during labour. When full eclampsia sets in, the mother's mouth twitches and her body contracts, then becomes completely rigid; violent muscular spasms break out and the woman foams at the mouth. So alarming is this complication that it probably prompted the first Caesarean sections to be conducted about two thousand years ago. Unfortunately, eclampsia remains a serious, potentially fatal condition, and can harm a woman's kidneys, liver, and blood vessels.

These terrible complications, however, do not affect all animals with placentas. In fact, they are only known to happen in three species alive today: patas monkeys, lowland gorillas, and us.

What distinguishes these three primates from other mammals is the extent to which the placenta penetrates the mother's blood supply. If the placenta does not invade deep enough, the mother's heart has to work harder, increasing the pressure of the blood in order to keep the foetus alive.

◉⊙◎

The link between pre-eclampsia and high blood pressure has been acknowledged since 1896, when the inflatable arm-band for measuring blood pressure was invented, but doctors still do not know exactly why, in some women, the placenta stops receiving blood as it should. The only risk factor that is universally accepted is being pregnant for the first time. Why should that be?

Our immune systems evolved to protect us from a staggering variety of parasites – anything that is in our bodies that shouldn't be. Once an outsider is recognized, the body's aim is to get rid of it. But we have seen that evolution has worked around this line of defence in many ways, for the simple reason that if a species is reproducing via sex, it benefits the mother's genes to become pregnant. Yet, it isn't easy for foreign sperm to get to an egg. Out of the three hundred million sperm that might be released into a woman's vagina, only one, if any, will normally succeed in fertilizing an egg. All the barriers are in place to prevent it: to prevent infection, the woman's vagina has an acidic pH that is also a killer for sperm; to stop microbes from invading, the cervix is filled with compact mucus, which also makes it incredibly difficult for sperm to make it to the womb; and then, the womb is armed with the soldiers of the immune system, white blood cells, which will physically engulf and destroy unwanted invaders, including most sperm. In pre-eclampsia, it may be that the woman's immune system has put up yet another line of defence and refuses to accept the incursion of the placenta's foreign DNA.

Pre-eclampsia occurs mostly in first-time pregnancies, but not all first-time pregnancies are the same. If a couple has had unprotected sex for less than four months before conception, the rate of pre-eclampsia is approximately four out of five. This decreases to one in four among those couples who have been having unprotected sex for five to eight months, and further to one in twenty among those who have been doing so for more than twelve months. Even a woman who has already conceived several children runs a heightened first-time-pregnancy risk when she takes a new male partner who isn't the father of her earlier children. Put simply, it may be that being exposed to a particular partner's sperm 'acclimatizes' a woman's immune system to his genes, breaking down the defences against foreign intruders and improving the negotiations that take place between the womb and the child. Becoming tolerant of a partner's sperm appears to protect the mother and the embryo, once it implants in the woman's womb.

So while humans may not ourselves be strictly monogamous, evolution has built women to have more successful pregnancies with long-term sexual partners.

◎◎◎

During pregnancy, the mother, too, has an active role in protecting the foetus from her immune system's attackers: substantial numbers of a woman's immune cells cross through the placenta and settle in the developing lymph nodes of the foetus, disguising the baby's immune system from her own. In a way, the body is tricked into seeing the foetus as a 'temporary self'. These cells also serve to suppress the foetus's immune system, which could be set against the mother's blood.

These maternal immune cells have an incredibly long-lasting influence on the foetus, even long after the baby is born. As they cross from the mother to the child, the cells 'teach' the foetus how to balance the need for self-defence against the need for tolerance to the surrounding environment. This is a tricky balancing act. If the foetus is taught to be too tolerant, the newborn baby may be left unprotected from a common but potentially lethal infection. If, on the other hand, the foetus's self-defence mechanisms become too keen, a child may be overly sensitive to certain foods and environments; worse, the body might start attacking itself – a condition called autoimmunity. Indeed, until at least early adulthood, a mother's immune cells influence how her child's body regulates its own defences and how tolerant, or susceptible, it will be to allergies and infection.

It used to be thought that a baby in the womb somehow made itself completely invisible to the mother's immune system, but this isn't strictly true. What happens, instead, is that the immune systems become interlocked. This means that diseases which are not normally transmissible between two adults can pass from mother to child. Diseases such as cancer.

In 2007, a twenty-eight-year-old Japanese woman gave birth to a girl. The pregnancy was uneventful, and the baby seemed perfectly healthy on delivery. When the baby was about one month old, however, the mother had to be rushed back into hospital: she was bleeding uncontrollably from her vagina. She died not long after being admitted. Although she had not known it, the mother had leukaemia, a cancer of the bone marrow and white blood cells, which is known to be a possible underlying cause of haemorrhaging after giving birth. Eleven months later, doctors found a huge tumour trapped in the baby's cheek.

Cancer cells tend to be pretty well skilled at making

themselves invisible to the immune system. Mostly, this is because they are actually our own cells, not foreign invaders, and the disease comes from mutations that make the cells incapable of regulating their own growth. They divide and spread and expand in ways that would usually mark them for self-destruction. But they grow on. That doesn't explain, however, why the mother's cancer cells would not be attacked by the baby's immune system.

When the doctors studied the cancer cells in the Japanese baby and samples taken from her deceased mother, they found that the cancer cells were missing a large chunk of DNA from chromosome 6. It is along this stretch of chromosome that the DNA normally produces the markers on to which our immune cells latch. In this case, the cancer cells passed from mother to child because the immune cells were not able to attack, and there was nothing the baby – no matter how vigilant her immune system might be – could do about it.

Inside the watery world of the womb, the growing baby receives many cues that affect its health. The body is primed in the womb for the environment in which the mother already lives. That, in turn, should increase the child's own chances of growing to adulthood and reproducing successfully.

For example, if a mother eats too much or goes hungry, the foetus will adjust its nutritional needs in both the short and long term, preparing itself for a world in which it will either have easy access to food or need to be ready to go without. Among the most well-publicized studies of nutrition in the womb are studies of rodents that have had their diets restricted. In particular, scientists have been interested to find out how a mother's calorie intake affects her fertility and the survival of her young. You might guess that a hungry mother rat would have fewer nutrients to share with its foetuses, and that less food would mean less offspring. This isn't the case. When female rats were fed a calorie-restricted diet, the mothers enjoyed a longer span of fertility, giving birth to pups at more advanced ages. And when these dieting rats gave birth, the survival rates of their pups were dramatically better than for the offspring of rats that were allowed to eat to their heart's content. For mother rats whose calorie intake was moderately restricted, over seventy-three percent of pups survived, whereas only twenty-two percent of pups survived that were born to mothers with an unrestricted diet.

Like the propensity to be allergic, however, humans are also programmed in ways that can make us oversensitive to certain chemicals, putting us at risk for related diseases. Coronary heart disease may also have its origins in the womb. Pregnant women who have impulsive, uncontrollable outbursts of temper (more incidents of slamming doors; loud, angry shouting; binge eating or drinking; smashing dishes; etc.) secrete higher levels of stress hormones, such as cortisol, which can cross through the placenta and reach the baby in the womb. Once there, the hormones change the way in which the *hypothalamic-pituitary-adrenal axis*, or HPA, and the autonomic nervous system work, and both the HPA and this part of the nervous system appear to be important for programming disease into the foetus. Individuals who were overexposed to stress hormones in the womb exhibit long-term, stress-related behaviour as adults. These hormones also affect foetal heart development and may increase the risk for developing cardiovascular disease later in life. Being overweight is associated with the release of inflammatory factors in the body, and these factors can also affect the development of the lungs and the immune system in a foetus. So if a mother is overweight, her child may also have a higher risk of developing allergies and asthma.

Obesity, in particular, may be decided in the womb – long before a child gets around to putting anything into his or her mouth. If a mother gains excessive weight or has diabetes while she is pregnant, the foetus will adapt to an environment where there is an excess of sugar around. As a teenager, her child is more likely to have a high body mass index, or BMI, even if the child does not eat fatty foods. Mice or rats that are put on a high-fat and high-sugar diet that makes them obese have pups that grow up to have increased body fat and abnormally large appetites. In fact, even when these pups are kept on a healthy diet, their appetites mean they are far more likely to become obese on standard meals. They also have an abnormally high level in the blood of the protein leptin, which has a starring role in the way food is consumed and then metabolized into energy by the body. Other experiments on animals indicate that if a mother's nutrition becomes imbalanced during pregnancy and breastfeeding, this permanently changes how – at the level of the brain – her offspring consume food as adults. The mother's patterns of consumption actually alter the developing hypothalamus, the almond-sized part of the brain that controls basic biological functions such as hunger, thirst, fatigue, and temperature. The hypothalamus produces the hormones that control when we feel hungry and desire food, as well as those that control aggression and sexual behaviour.

◉◉◉

Bizarrely, because of how imprinting works, a father also sways his child's life after birth, through the timed influence of his genes. Some of a father's genes, in fact, only become expressed after a baby has been weaned, and they can have a significant effect at much later stages of life. For instance, the father's genes have a say in whether or not a child develops disorders related to food, possibly including obesity. Here, once again, we see the battle between the father's DNA and the mother's body.

This is because those eighty genes that are subject to imprinting have an important say in the development of our brains. Those genes that are silenced when inherited from the mother but expressed when inherited from the father inhibit our overall brain size; they contribute to the development of the hypothalamus – the impulse centre that makes us crave food. In contrast, the imprinted genes that are expressed when inherited from the mother contribute to the cortex, the so-called grey matter of higher mental functions; the striatum, which is involved in decision-making and risk-taking; and the hippocampus, the brain's memory centre. Recent molecular analysis has shown that among people who carry defects or mutations in genes that are supposed to be imprinted, there is a surprisingly large incidence of cognitive, behavioural, neurological, and psychiatric conditions. These include autism, bipolar affective disorder, epilepsy, schizophrenia, and Tourette's syndrome.

This sex divide in the role of imprinted genes on the brain is curious, because what happens in the hypothalamus is also believed to influence maternal behaviour. Studies in mice have shown that mothers will neglect their offspring if the *PEG1* gene (paternally expressed gene 1) is removed from their fathers so that they are not able to inherit it. A related gene, called *PEG3*, increases maternal care, too, and also regulates male sexual behaviour – meaning it ensures its own preservation. If a male mouse does not have the *PEG3* gene, then, no matter how much sexual experience it gains, it is unable to improve its reproductive effectiveness; for example, no amount of experience will make the mouse better able to recognize the odours that female mice secrete when they are ready to mate. It seems that a father can even directly influence how his daughter will behave towards her own children, through imprinted genes alone.

For mice and men (and women), evolution has pitched mother against father, father against mother, mother against child, and child against mother – our genetic sources and our genetic creations are all battling for control. The outcome of these long-ago skirmishes is a treaty written in DNA: neither a mother nor a father may use all the genes at their disposal, but both will have a genetic and a chemical voice that will continue whispering into the brains of their children – all the way into adulthood. Neither sex can do without the other. At least, that is, while we are still constrained by the body.

But what comes next? Will those restrictions still hold when eggs and sperm and foetuses and wombs are no longer tied to biological packages of human anatomy? Because, if the past centuries of discovery show anything, it is that once science stumbles upon an obstacle, the next step is to tear that obstacle to bits, find out how it works, and then see if we might get rid of it altogether.

PART II

A NEW WAY OF MAKING BABIES

All our science is just a cookery book, with an orthodox theory of cooking that nobody's allowed to question, and a list of recipes that mustn't be added to except by special permission from the head cook.

Aldous Huxley

6

OUT OF THE TEST TUBE

If every physical and chemical invention is a blasphemy, every biological invention is a perversion... And all this of course applies much more strongly to the sexual act.

J. B. S. Haldane, *Daedalus*, 1923

On 27 July 1978, the front page of the London *Evening News* carried a nearly life-sized photograph of a beautiful infant, a mere eighteen hours old. She was wide-eyed and swaddled all in white, a perfect specimen of just-born humanity, and she was pictured below the headline – SUPERBABE – that announced her birth.

Louise Joy Brown was no ordinary newborn, but 'the world's first test-tube arrival', a child widely celebrated as a miracle of science. As her middle name testified, Louise was an even greater miracle to her infertile mother, thirty-year-old Lesley Brown, whose grossly distorted and persistently blocked Fallopian tubes had made it impossible for her to become pregnant over the course of nine arduous and depressing years. Louise was the first child to be born through *in vitro fertilization*, or IVF.

Medical science had been wrestling with infertility for quite some time. But because the subject has been (until very recently) shrouded in secrecy, it is almost impossible to say with any accuracy when artificial insemination in humans was first attempted by doctors. 'Historically, artificial insemination is one of those rare medical entities which cannot be traced back to Hippocrates,' wrote one American obstetrician back in 1943. Yet, we can trace the practice to at least the middle of the fifteenth century, when a French doctor called de Villeneuve performed artificial insemination for King Henry IV of Castile and his second wife, Joan of Portugal. The king was rather unkindly nicknamed 'The Impotent' – and although local prostitutes confessed to a priest that their monarch was perfectly sexually capable, close examination confirmed that he could not, indeed, get an erection. Artificial insemination, however, was not successful. De Villeneuve could not have known that King Henry was probably living with a pituitary tumour or a condition known as hypogonadism, either of which would have rendered him completely sterile. So whatever ejaculate he could supply to his doctor contained little or no sperm. Even if de Villeneuve had managed to introduce the royal semen into Joan's womb, pregnancy would have been near impossible. Fortuitously, his wife took matters into her own hands and bore three children by natural donor insemination – that is to say, the children were, reportedly, fathered in amorous liaisons with the Duke of Albuquerque and with the nephew of a Church bishop.

The first authenticated case of artificial insemination was performed successfully about three hundred years later, around 1776. In this revolutionary year the renowned surgeon and human dissector John Hunter was approached by a 'linen draper in the Strand' of central London who came to consult him because of a deformity of his penis. The draper suffered from hypospadias, in which the opening of his urethra, from which a man would normally ejaculate semen, was in a position on his penis from which his sperm could not physically make it into his wife – the same condition that likely afflicted King Henry II (and his bride Catherine de Medici). Hunter armed his patient with a syringe, and advised him to use it to collect his semen after sex and inject its contents directly into his wife, while she would be most receptive. Presumably Hunter felt she would be more open to having a syringe inserted into her vagina around that time, but he may also have suspected that this moment would give the semen (and its sperm) a better chance of making its way well into the uterus. It was a simple but effective intervention. The wife became pregnant, and the baby was born healthy.

Such hit-or-miss attempts were always in demand, though they were not always scrupulous. We have records of a French doctor by the name of Girault, who in 1838 used a hollow tube to blow sperm into the vaginas of infertile men's wives. Another French doctor was forced by public disapproval to cease similar attempts to impregnate women artificially. After the American Civil War, James Marion Sims – credited as the father of American gynaecology – reported his own attempts at artificial insemination, in which he injected sperm past the vagina, directly into the womb.

Sims had a reputation for medical miracles: he was also known for curing crossed eyes, clubbed feet, and a debilitating condition called vesico-vaginal fistulae (VVF), which affects women. VVF is a trauma commonly associated with a prolonged, obstructed labour, during which the baby's head puts pressure on the tissue that normally forms a barrier between the mother's vagina and her bladder. If the baby gets stuck and remains in this position too long, this tissue can be destroyed, and a hole opens between the vagina and the bladder. This often leads to constant, uncontrollable urinary incontinence – a debilitating situation, physically and emotionally. In Sims's day, women with the condition were likely to become social outcasts. In his efforts to find a surgical solution to this problem, between 1845 and 1849 he carried out a series of operations on black slave women. In that day, any woman with VVF would probably have accepted the slimmest of chances to be rid of it, so the slaves he operated on may well have been consenting; however, anaesthesia had only recently been discovered, and some accounts say he performed what must have been incredibly painful procedures on the slaves without the anaesthetic that he later used with his white patients. The alleged practice has left his medical legacy in something of an ethical limbo.

Regardless, Sims published his definitive work on women's reproduction and 'uterine surgery' in 1866. In it, he also logged the fifty-five artificial inseminations he had conducted on six different patients using sperm from their husbands. Bypassing the vagina to put seminal fluid directly into the womb was excruciating for the patient; Sims himself states that his earliest insemination experiments were 'often more painful than any operation'. Half of his attempts he considered to be utter failures, and only once did he achieve a pregnancy. Sims's poor results probably had less to do with his technique than with the era's limited knowledge about menstruation, and about where in a woman's body conception actually happened. It was truly a guessing game. Unfortunately, the only time Sims guessed right and managed a successful artificial insemination, the woman miscarried, having experienced a 'fall and a fright' when she was four months into her pregnancy. The twenty-eight-year-old patient had undergone Sims's procedure ten times. After that twist of fate, the doctor wrote, he gave up the practice altogether.

Other doctors forged ahead, however. To test the limits of artificial insemination, they soon began to turn to donors, breaking from the tradition of exclusively using sperm from a woman's husband. The world's first case of such donor insemination was performed in 1884 by Professor William Pancoast, who was based at Philadelphia's Jefferson Medical College. Pancoast used a hard rubber syringe to insert sperm donated by one of his medical students, whom he had judged to be the best-looking of the bunch. His patient, a woman who had been anaesthetized prior to the insemination, was unaware that Pancoast had even performed the procedure, and then, when her infertile husband was made aware of Pancoast's procedure, the doctor instructed him never to tell her about the day's events. Their son was never told the circumstances of his birth either. Only the medical archives have given the story to us.

At the time, the thought of using 'alien' semen shocked many people. In the nineteenth century, the idea of sperm banks had inflamed the imagination of doctors and the public alike; as early as 1870, there was speculation that soon there would be places where you could buy the semen of a 'thirty-year-old blond with black eyes' or a nineteen-year-old virgin. These were not considered to be happy developments. This attitude persisted for decades – at least until the 1940s in the US, and into the 1970s in Australia, and was only supplanted after the first professional sperm banks were launched in these countries. Indeed, in Pancoast's day, there were still so many dissenters opposed to impregnating a woman using sperm that was not her husband's that he kept his work secret until his death, and it was only revealed in 1909. In fact, the obituary for him, printed in the *New York Times* on 6 January 1897, made no mention of his innovations in this field. Instead, he was remembered as the surgeon who performed the autopsy of Chang and Eng Bunker, the conjoined 'Siamese twins' who had been exhibited in Victorian freak shows nearly seventy years earlier.

A contemporary of Pancoast's working in Austria, Professor Leopold Schenck, decided to go one better. In a Petri dish, he mixed together the sperm and egg of rabbits in an attempt at developing a rabbit embryo – something close to in vitro fertilization proper. But Schenck never succeeded in making bunnies, let alone babies. His clinic instead became renowned for a different speciality: that he could influence the sex of a baby, as its parents desired. And for this, he was in great demand. So much so that in 1898, a visiting American doctor barely gained entrance to Schenck's Parisian surgery. On a trip to Paris, Dr Victor Neesen, who was visiting from Brooklyn, noted, 'When I called at Dr Schenck's house I found the street blocked with carriages of all descriptions. A group of well-dressed people stood on the stoop of the house, waiting to be admitted. The anterooms were crowded to suffocation with visitors, most of them women, richly attired and genteel looking, all waiting to consult the professor.'

Though there were those who objected on moral grounds, there was evidently quite a strong appetite for all sorts of reproductive manipulation. As long as it happened inside the body, and not in a test tube.

◎◎◎

Not surprisingly then, there was huge public interest in the first IVF baby. Before the birth of Louise Joy Brown was announced in the summer of 1978, the Oldham and District General Hospital enlisted a guard dog as backup to its usual entourage of security guards, to ensure that mum and baby could avoid the paparazzi and get some much-needed rest. Still, journalists besieged the hospital's maternity unit. It was suspected that the tabloids may have even faked a bomb scare, in efforts to snap a photo of an exhausted Lesley Brown as she left the building.

There had been a long wait for the child that many referred

to as 'our' baby. The animal research into the potential for IVF had been encouraging, but there were fears that the process would not work in humans, or that terrible abnormalities would result. When Lesley and John Brown arrived in the office of Dr Patrick Steptoe at Oldham General Hospital outside of Manchester, they had nearly given up hope. Steptoe and the Cambridge physiologist Robert Edwards had been investigating options for fertilizing eggs outside the womb since the mid-1960s, but the eighty pregnancies they had managed had only lasted a few weeks before spontaneously aborting. After extracting an egg from Lesley and fertilizing it with John's sperm, they watched it divide into more cells – and they took a gamble. Rather than waiting four or five days, as they had in the past, they injected the fertilized egg into Lesley's womb two days later. The egg attached itself to the uterus wall without any difficulties.

Once news of the pregnancy leaked to the media, Lesley was forced into hiding. The press were chasing her all over Bristol, where she lived, and Edwards and Steptoe were concerned that she might lose the baby from the stress. Eventually, Steptoe drove Lesley to his mother's house in Lincoln. For the rest of the pregnancy, the press could not uncover where she was.

For weeks, things seemed to go well. Then, near to her due date, Lesley's blood pressure spiked. The doctors chose to deliver the baby early, rather than risk complications from natural labour. 'There were many times in the last ten years when we wondered if we would ever see that baby,' one of the team of doctors who had been working with Steptoe and Edwards on IVF later told the BBC. On the night of Louise's delivery, there was a buzz in the air. Some of the medical staff had even drawn lots to be present at the history-making birth.

Louise may have been the first 'superbabe', but she was not to be the only one – even as she was born, there were reports of other mothers who were already pregnant using the new technology. By 1979, two more 'test tube' babies were born in the UK. The next year, Australia greeted its first IVF birth, with the United States matching this feat the year after that. Between 1978 and 1999, fifty thousand IVF babies were born in the UK, and more than four million worldwide. Each year, around eleven thousand IVF babies are now born in the UK, and around forty thousand in the US.

It may seem strange, but the world's first IVF baby was also a spur to much ideological controversy. As with some views about human cloning today, many then (as some still do now) felt that creating life 'in a test tube' was unnatural; they believed that conception was supposed to happen through biological sex – and sex between married partners, to be quite precise. Among those opposing IVF was the Catholic Church. 'The fact that science now has the ability to alter this does not mean that, morally speaking, it has the right to do so,' the general secretary of the National Conference of Catholic Bishops told the *Washington Post*. But reaction was mixed, and the Church's sentiment was not wholly embraced by the public. A Gallup poll taken shortly after Louise's birth attested that sixty percent of people favoured the new technology, because it would make children possible for those couples who would otherwise be unable to have them – a clear majority, but not an overwhelming one.

Yet, for the five thousand or so childless couples who promptly signed up for the new technology, IVF couldn't have come soon enough. Here was a dramatic addition to the arsenal against infertility, one that could make them parents, even if biology was against them.

◎◎◎

Today, *assisted reproductive technology*, or ART, involves far more than IVF, and includes any procedure in which eggs, sperm, or embryos are manipulated in vitro. Around one in six couples will seek medical assistance because they are having difficulty conceiving a child, and only about one in ten who consult a doctor will go on to use IVF. This is because, though it is probably the most widely discussed reproductive technology, IVF is not the only, or the first, choice when a problem with fertility is diagnosed.

For instance, many couples will turn to induced ovulation, where a woman is treated with hormones to stimulate her ovaries to release eggs. While the ovaries are being manipulated through biochemistry, a couple may conceive merely by timing their sex; they know when the woman's egg *should* be triggered for release by the hormone soup she's taking. That's much better than guessing.

For couples with unexplained infertility, induced ovulation is sometimes accompanied with *intra uterine insemination*, or IUI, which can help to get underperforming sperm or eggs in the same place, at the same time. In IUI, semen is first processed in the lab, so that only sperm that are moving can be selected. These 'good' – or, at least, somewhat mobile – sperm are then delivered directly into the womb, near where the egg is preparing itself to be fertilized, bypassing the hazards that might assault them in the vagina and cervix, namely, unfavourable pH environment and tricky-to-navigate mucus. The vast majority of sperm are normally killed somewhere en route. Among couples with no clear fertility dysfunction, there seems to be a significantly higher chance of a successful pregnancy when IUI is used with ovarian stimulation rather than on its own, although simply having sex at the right time, which is much easier to pinpoint with ovarian stimulation, might work just as well.

It may also be the case, as with Lesley Brown, that the reason a woman cannot get pregnant is because there is a blockage in her Fallopian tubes, the two passages that connect the ovaries to the womb. If this is the problem, a woman might choose to have surgery to unblock the tubes. But if this fails, or if the woman is already in her mid- to late thirties, when the biological clock is ticking away, IVF may indeed be the best bet.

In IVF, an egg is extracted from a woman and incubated with around fifty thousand sperm. The in vitro part of IVF literally means 'within glass', because it originally referred to the fact that the experiments were performed in glassware, which was commonly used in labs instead of plastic. (This is in opposition to 'in vivo' fertilization, which comes from the Latin for 'within the living'.) Fertilization in vitro thus means that the egg and sperm are effectively left to mingle and 'introduce themselves' in a specially created, artificial environment, something like a speed-dating night. With the plethora of obstacles in its way safely removed, a sperm can make a direct hit on the egg – as long as it can beat its competitors to it. It follows from the particulars of the set-up that for men whose sperm cannot move properly to the egg or penetrate it to begin fertilization, IVF may simply not work.

By the late 1980s, fertility doctors had found a way to solve this matchmaking problem, without having to resort to donor sperm. Several techniques emerged that could improve an infertile man's chances of making a baby, with one frontrunner quickly establishing itself: *intra-cytoplasmic sperm injection*, or ICSI. Like IVF, ICSI does not correct defects in the sperm, but it requires a much smaller number of sperm: instead of fifty thousand sperm, just one will do. ICSI does not leave fertilization to chance. This single sperm – possibly immobile, but otherwise picked for its 'good looks' – is first stunned, for example by rubbing its tail. It is next sucked up, tail first, into a sharp-tipped glass pipette and injected directly into an egg. Or, instead of a whole sperm, the doctor can pluck just a single sperm head or nucleus, containing the DNA that provides all the necessary genetic instructions for making a baby – the rest of the sperm is simply a vehicle to get the male DNA into an egg.

Neither IVF nor ICSI are without limitations – their success rates are, at best, only around thirty percent. These remain techniques for assisting natural reproduction, not for replacing it.

But as the use of ART increases worldwide, it is certainly also important to consider some potential long-term consequences. For a start, in both IVF and ICSI, eggs need to be removed from a woman in order to ensure sperm access to them. But eggs are tucked away in the ovaries, and released only infrequently following a biological programme, which makes them far trickier to acquire than sperm. To harvest her eggs, the doctor places the patient under local anaesthetic, and then, guided by ultrasound, passes a long, thin needle through her vagina to the ovaries. The needle is used to suck the fluid out of mature follicles, or egg-containing sacs. If an egg happens to be retrieved with this fluid, it is gingerly removed and tucked into an incubator. And then the process is repeated, until several more eggs join it.

This process of extracting eggs is not just invasive; it carries very real health risks as well. To persuade the ovaries to release multiple eggs from their grip, these organs must be stimulated with a suite of hormones that the body itself uses for that purpose, but often at higher than normal doses. These include a drug to stop eggs from being released until they are mature (a *gonadotropin-releasing hormone*, or GnRH, agonist or antagonist; a *follicle-stimulating hormone*, or FSH, that kick-starts the development of multiple eggs; and *human chorionic gonatotropin hormone*, or HCG, which forces the eggs to mature properly (and is the hormone that causes over-the-counter tests to test positive for pregnancy). Without stimulating medications, the ovaries generally produce one egg a month. With them, they churn out anywhere from five to twenty-five eggs that might be harvested, and some young women have been reported to produce fifty to seventy eggs from one high-dose stimulation. This sounds like good news, but this stimulation can cause severe *ovarian hyperstimulation syndrome*, or OHSS – most likely as a result of too high a dose of HCG. When OHSS sets in, the woman's blood vessels become much more permeable than they should be. This leads to a drop in blood volume, so that her blood thickens. It can lead to organ failure, and is a life-threatening condition.

The cocktail of medications served up in IVF and ICSI can also cause developmental problems in the babies they beget. FSH, if given in high doses, may lead to the production of eggs with too many or too few chromosomes or of eggs with chromosomal abnormalities. Having the right number of chromosomes is incredibly important; for example, in Down syndrome three copies of some genes are carried on chromosome number 21, rather than the usual two copies (an issue we will look at in more detail in the next chapter). High levels of FSH might also have a detrimental effect on the lining of the womb, which could compromise the growth and health of the baby growing in it. Children conceived in vitro, and even those conceived merely with the assistance of ovary-stimulating drugs, may also be more likely to be born with several syndromes. Babies with Beckwith-Wiedemann syndrome grow too large – both in the womb and outside it. This overgrowth can affect all systems of the body and cause diabetes, abdominal wall defects, kidney problems, and embryonal tumours. Another condition, Angelman syndrome, may express itself through severe mental retardation or delayed motor development, which means poor balance, jerky movements, and difficulties with speech. The growth disorder Silver-Russell syndrome leads to dwarfism. Fortunately, all of these conditions are rare in the general population, so an elevated risk after IVF or ICSI does not translate into an epidemic among scientifically fertilized offspring. Indeed, the absolute risk for a serious congenital malformation or chromosomal abnormality after IVF and ICSI appears to be small.

What is interesting from a medical perspective is that the last three syndromes are model 'imprinting' disorders – they result from changes in genes that work selectively depending on which parent provided them. Typically, of course, we inherit two complete sets of chromosomes, one from our mother and one from our father, and most genes are an expression of both of these sets. Imprinted genes, however, are expressed from only one of the pair of genes, the mother's or the father's, and as we saw earlier, many imprinted genes are critical in normal growth and development – hence the problems here with overgrowth and mental retardation.

Around half of all children diagnosed with Beckwith-Wiedemann syndrome have lost the key to a gene inherited from their mothers, and this change has been detected in almost one hundred percent of children with the syndrome who were conceived after any brand of ART. Experiments with laboratory animals indicate that imprinting disorders that occur as a result of IVF may be triggered by a few specific techniques – and so the risk might vary from clinic to clinic.

Remember that genetic imprinting was only discovered in 1984 – and the first child was born using IVF in 1978. For that reason alone, imprinting defects in IVF babies could not have been predicted from the start. Since then, reams of research have been compiled to compare the genetics of naturally- and laboratory-conceived children. But why should children born through ART be more susceptible to imprinting disorders? What is it about manipulating egg and sperm that leaves genes unable to 'tell' which parent they hail from, and whether and how they are supposed to go to work on developing the foetus?

◎◎◎

Thus far, geneticists have learned that these imprints are erased in the cells that are specifically programmed to become eggs or sperm – a tantalizing piece towards solving the puzzle. While eggs and sperm are being made, the imprint is reset accordingly – in sperm, a certain subset of genes will be rendered non-functional; in eggs, a different set. It's a reversible process that depends on the parent of origin, and determines the different functions of the eggs and sperm as they are developing. This means that the very process of making eggs and sperm is critical for the 'right' genetic imprinting, and ART procedures affect these developmental periods when genomic imprints are so vulnerable.

Genes are imprinted much earlier during the production of sperm than they are for eggs. For this reason, if you were to induce an egg to mature artificially, as happens in induced ovulation, you might disturb the genetic imprints that should be taking place in the egg, but are unlikely to affect the sperm. In fact, for growing eggs, the genetic imprints are not completed for some genes until just prior to ovulation. If this vital process is vulnerable to the hormones used, then it's not surprising that in these developmental syndromes, the fault always lies with the egg. But even after fertilization has occurred, there is another critical period for getting the embryo's genetic imprints right. In ART, it happens that embryos are usually still in vitro during this second, vital window, making them vulnerable once again.

Further, infertility is often linked to genetics, and these genetic problems may be inherited by a child produced through ART. This raises the provocative question of whether generations of babies created with IVF or ICSI might 'naturally' pass along genetic defects that will lead to a significantly more infertile population. In December 2006, Louise Joy Brown, our first test-tube baby, gave birth to a healthy son; seven years earlier, Louise's little sister Natalie, who herself was the fortieth child born by IVF, became the first child of ART to have a baby of her own. Neither of the Brown sisters had required fertility treatment. Of course, Lesley and John Brown had turned to IVF because of a blockage in Lesley's Fallopian tubes, a condition that can often be corrected with surgery. If it had been John whose fertility had been the main issue, and if Louise and Natalie had been a Louis and a Nathan instead, then the next generation might have been more complicated to conceive.

This is especially true for patients of ICSI, which is used when sperm is abnormal. Although ICSI takes longer and is more invasive than artificial insemination using donor sperm, couples trying to conceive still tend to prefer using their own abnormal sperm; a child who is genetically 'their own' outweighs all other considerations. Chromosome anomalies are seen in about seven percent of men who fail to produce sufficient sperm – and among this seven percent, more than ninety-nine percent of whatever sperm they do make will exhibit abnormalities.

There are concerns about the effect of using abnormal sperm for ICSI, because abnormal sperm are associated with increased chromosome defects in the babies produced. The chance of having a baby with major malformations through ICSI is twice as high as in the general population – nine percent, versus three to four percent. This may be because abnormal sperm also tend to carry the wrong number of chromosomes. In men with very low sperm counts, seventy percent or more of their sperm will carry too many or too few chromosomes. Moreover, the most commonly recognized genetic cause of infertility in men is the appearance of Y chromosomes with corrupt or missing genes. The genes on the Y chromosome are essential for sperm production – this is, after all, the chromosome that makes males male. But these missing genes could hint that there are abnormalities on other chromosomes too. Indeed, there is evidence that ICSI children have an increased number of abnormalities, mostly inherited from the father's side, and that ICSI sons are more likely to be affected than daughters.

The missing genetic material could make many of these sons infertile. Already, adult men whose mothers received fertility treatment are reported to have lower sperm concentration and count, more abnormal spermatozoa, smaller testes, and lower testosterone levels. Boys conceived by ICSI sometimes have reduced levels of testosterone. And ART has been associated with hypospadias and another condition, cryptorchidism, where one or both testicles fail to move down into the scrotum before birth. Most of these cases do resolve on their own, but sometimes surgery is required. Unfortunately, as incidents of hypospadias and cryptorchidism increase, poor semen quality and the rate of testicular cancer rise too. So boys diagnosed with hypospadias or cryptorchidism will need to be monitored for testicular cancer throughout life. IVF and ICSI also increase the chance of preterm birth, low birth weight, and multiple births. And in premature boys, an undescended testicle is more common.

Studies also suggest that Y chromosomes with DNA deleted in particular regions may cause babies to be born with two X chromosomes in some cells of their bodies, but with only one X chromosome in others. While this only impacts the sex chromosomes – the X and the Y – and does not, it should be said, create the situation seen in baby FD or in Jane from Boston, it does result in baby girls who have sexual ambiguities or Turner's syndrome, which creates females without female sexual characteristics. People with Turner's syndrome are noted for their short stature, and for a likelihood of other health problems, including difficulties with hearing, sight, thyroid and kidney function, high blood pressure, diabetes, and learning. Their ovaries also don't work – so these children are nearly always, like their fathers before them, infertile. (They may also, like post-menopausal women, suffer from osteoporosis because of their failed ovaries.)

The bottom line is that, when making babies through ICSI, you are often working with screwed-up sperm. The more screwed up the sperm are, the more abnormalities you will see in them, and the more likely they are to carry damaged DNA. And because the Y chromosome is only ever passed from father to son, if a man is infertile because parts of his Y chromosome are missing, his ICSI son, by definition, will inherit that corrupted Y chromosome and his infertility too.

◉◉◉

On the other hand, IVF and other forms of ART also now make it possible to diagnose genetic abnormalities very early – by the time the embryo is three days old – something that is not possible in natural pregnancies. As our techniques improve, screening an embryo to identify an abnormality before it is implanted in the mother's womb could be utilized to reduce further and further the chances of the diseases that are made more likely by ART.

Today, there are two ways of approaching the process of screening. One, called *pre-implantation genetic diagnosis*, or PGD, looks for specific genetic disorders that a couple is known to be carrying, and which therefore they have a high risk of transmitting to their children. PGD takes three to four days, and is primarily used by fertile couples who are worried about a particular disease that runs in the family. The second is a less targeted technique, known as *pre-implantation genetic screening*, or PGS, which has a turnaround time of twenty-four hours. PGS looks for mistakes across all of an embryo's chromosomes, using tests that can detect any abnormalities in an embryo's chromosomes – for instance, if it gained extra chromosomes, or lost some of them. This technology is still new and evolving, but it is likely to improve in the next few years. As it does, the genetic weaknesses involved when IVF or ICSI is necessary will almost certainly become less of a problem. With better PGS, fertility doctors will gain the ability to pick the most normal embryos.

Of course, while genetic problems may be soon within the reach of science to resolve, IVF and ICSI also give birth to complex moral conundrums that would never arise in a world where every pregnancy happens through sex. When fertilization occurs outside of the womb, and the embryo is then placed there, a woman becomes able to carry a child to term who is not genetically her own. For a woman who does not have good quality eggs, this is a great advantage, because she can choose to use an egg donated from another woman.

The technology has also become a very efficient way for older women to have successful pregnancies – by freezing a number of eggs or early-stage embryos from which they can select, and then trying each one out. Some women choose to freeze their eggs at a young age, and use these healthier eggs later in life, when they are ready to have children. But mistakes do happen.

In 2009, Caroline Savage, a forty-year-old American mother of three, returned to the fertility clinic where she had previously received IVF – and got pregnant. The clinic had kept frozen five of her early-stage embryos, left over from her last cycle of treatments. Unfortunately, there was a mix-up, and the embryo implanted into her womb was not one of her own; it belonged to a completely different couple, who also had 'leftover' embryos stored at the clinic.

Ten days after the procedure, Savage received a call from her doctor, notifying her of the error – news she later described as the worst of her life. The clinic's directors offered her a choice: an abortion (free of charge, one presumes) or a surrogate pregnancy (after which she would give the child to its rightful genetic parents). Savage opted for the latter, on religious grounds, and because she realized that if one of her embryos had been mistakenly inserted into another woman's body, she would go to the ends of the earth to get back her child. And if that hypothetical surrogate had chosen the abortion, she would have been helpless to stop it. In the state of Ohio, where Savage lived, surrogacy agreements are open to interpretation, though genetic parents are considered natural and legal parents of a child that another woman has carried. This was no surrogacy case, however; there was no intention, let alone an agreement, to have someone else's baby end up in Savage's womb. Yet, in this case, Ohio law recognizes the woman whose womb the foetus is in to be the mother of the child, rather than the woman who is genetically related to it. As mere donors of genetic material used to create that embryo, the other couple have no parental rights or responsibilities with respect to the child being carried to term.

Wracked with this knowledge, Savage and her husband asked a lawyer to reach out to the genetic parents, and three months later the couples met. The pregnancy was a difficult one for Savage, and she was scheduled for a Caesarean section. She cannot now risk another pregnancy herself, but still wants to grant a chance of life to her remaining embryos. To do so, she will have to hire another woman to carry the embryos to term. Savage would never have given birth to someone else's genetic child in a world without IVF, but nor would another woman have been able to give birth to hers.

◎◎◎

Infertility is a complex problem with many causes, and its solutions present just as many ethical conundrums. A century ago, European doctors tried to allay the public's fears by claiming that there was nothing truly 'artificial' about this new method of insemination. After all, the babies produced would be very real, the equal of any who had been naturally conceived. The field of reproductive medicine was simply a way of assisting nature.

Today, around one out of every fifty babies born in the UK, and one in a hundred babies in the US, starts life in a lab. What is more, starting life in vitro is no longer seen to be unnatural. In Europe, around one in four young men now have a sperm count that would render them subfertile or infertile; they will likely need to use ICSI if and when they decide to reproduce. An estimated sixty thousand women in Britain seek IVF every year. By the current medical definition of infertility – the failure to achieve a pregnancy within one year of regular, unprotected intercourse – some nine million people in the UK fall into this category.

There are many women with abnormally shaped wombs, unhealthy eggs, or no eggs; many men whose sperm are just not up to scratch; and men and women who have had, for example, treatments for cancer that have killed off their reproductive material before they have had a chance to become a parent. Some couples can't have children because one of them is infertile, but if a couple cannot have a child because they are two men or two women, then technically, they are infertile too.

Bringing egg and sperm together cannot, by itself, resolve all of the issues that people may face when they want to have a child but cannot. So it makes sense that, one day, possibly soon, we will expand our means of reproduction to be far broader than our current repertoire. To get a glimpse at the future of reproduction, simply think about the problems that ART has not yet resolved. Who are the Lesley and John Brown of the next phase of human history?

Thirty years after the first test-tube baby, science is poised to add many new weapons to its armoury in the battle against infertility, including using your own stem cells to generate fresh sperm or eggs or both, when you don't have any or have run out. There may perhaps be gene therapy to prevent miscarriages from corrupt chromosomes. There is even a body of research to prepare us for reproduction in space, where sperm seem to move faster (a fertility plus) but some hormones may not be activated (a developmental negative).

Since humans first evolved, men and women have needed each other to make babies. But the nature of human reproduction is about to change radically. Children born this year will be able to make babies in ways their parents could barely dream of – when, that is, they decide to have children, at a time *entirely* of their own choosing.

7

OUT OF TIME

There will be nothing but time, don't you understand?... If I can have a child at seventy-three, then why should you have one at forty-three, or forty-five?

Ann Patchett, *State of Wonder*, 2011

The Mosuo people live high in the Himalayas, in the Yunnan and Sichuan provinces, near the Tibetan border, in China. They live a primarily agrarian life, raising yak, water buffalo, sheep, and poultry. They are also one of the few peoples whose language appears to have no word for 'father' – perhaps the most exotic facet of their way of life, to an outsider.

The Mosuo are a matrilineal culture: it is the women who determine the family line and inherit the family property. Such practices have existed in Tibetan and northern Indian societies from Neolithic times, presumably motivated by a desire to keep wealth and resources within the kin group. These customs started to decline – or at least, to be hidden – after missionaries and colonists began to malign them in the nineteenth century. Among the Mosuo, however, maintaining a matrilineal culture grew into something of a necessity as more and more Mosuo men started to leave their villages to become monks or trade along the Silk Road.

With their men absent or unavailable, the Mosuo women took over the day-to-day administration of the community. They chose not to marry, opting instead to look after their own households, some populated with four generations of Mosuo women. At puberty, a girl would be given a private bedroom, in an otherwise open-plan home, and like a society debutante she would attend dances, looking for a suitable partner for courting. If a young man caught her fancy, the girl would be free to choose him as her lover – and not just as a lover, but as a father to one or more of her children. But their relationship was temporary; the man might be allowed to stay the night, but in the morning he would go back to his own dwelling, to live with his own mother's line. There was no requirement – no expectation – that the father would stick around.

These so-called walking marriages involve none of the messy elements of a long-term partnership. There is no divorce, no joint-property disputes, no custody battles. In this way, new generations would be born and the missing men would be replenished, with the village remaining peaceful under the watchful eyes of the mothers.

Walking marriages may appear to sanction promiscuity among young Mosuo women, but these arrangements are more like a form of serial monogamy, which is widely and happily practised across the Western world. The main difference here is that in a Mosuo village there is no stigma attached to single motherhood. This is the norm, and mothers continue to live with their extended families – grandmother, mother, cousins, aunts, and uncles – as they raise their own children. Uncles stand in for the father when it comes to providing a male role model. And this means that any woman can have a child *when* she chooses, without the stigma of single motherhood – something that many Western women cannot say of themselves.

⊚⊙⊚

There have always been cultural norms around reproduction – norms involving what religion, caste, or race a partner should belong to; whether polygamy or monogamy is acceptable; which children are top of the pecking order to inherit property; whether or not an elderly man is an acceptable mate for a younger woman. But a recent survey carried out in more than seventy countries shows that our values about reproduction are tied not just to our family's status and access to property, but also to the larger context of economic development. In industrialized societies, this encompasses a remaking of the idea of family that is completely changing the way we have babies – and what the future will look like.

At the most basic level is a change in our interactions with close family members, who are now often replaced with friends, colleagues, and peers in our social lives. This is especially true early in adulthood, when people enter the workforce and establish their own homes. At the same time, the level of education you're expected to have in order to get a job has increased, which means that people are staying at school, at university, and in training, and building up debt, into the early years of adulthood. Men and women contemplating a family of their own face a long list of obstacles: to be able to afford appropriate housing; to afford or clear university debts; to achieve job stability and a level of wages that can support more than one person (or two); and to make a career and childcare compatible. And that's setting aside the very modern desire to win the lottery of finding a 'good' partner, in romantic terms. With all of these obstacles standing in the way, having children is less of a priority – or, at least, less of an immediate one. So men and women in the industrialized world are increasingly waiting to start families until their mid-thirties, or even into their forties.

Global fertility is in general decline. This trend is most pronounced in industrialized countries, especially in Western Europe, where the population is projected to decline dramatically over the next fifty years. According to the US Census Bureau, nearly all the world's developed regions are reporting fewer births, and about half the world's population lives in regions where the number of births is fewer than necessary to achieve long-term population stability. And that's just in the short term. Of the 223 countries listed in the CIA's *World Fact Book*, ninety-four now post a fertility rate of less than two children per woman – a rate that means fewer kids to support ageing parents. The ninety-four countries on the CIA list include China, Japan, and South Korea, all of which have lower birth rates than do the countries of Europe. Birth rates are certainly falling in Europe and in the US due to the current trend towards starting a family later in life. The US rate squeaks above two, at 2.05 – putting it just outside this club. The European Union's statistics agency predicts that by 2050 the federation's population will drop by around seven million.

Studies show that seventy-four percent of women who definitely or probably want children in practice delay getting pregnant because of relationship issues. For most women, it's not work or training worries, or other distractions, that cause the delay, but basically because they hadn't yet found the 'right' partner. That's not to say that career ambitions don't play a role. The second most common cause for delayed motherhood in developed countries is the social rewards that come to those who achieve professional success. Things like educational status, a prestigious or lucrative career, and the allure of luxury (and sometimes not-so-luxury) goods interfere with a woman's opportunities to reproduce, because acquiring the trappings of success takes energy and time, right at the peak ages for fertility.

These sociological and economic shifts have had the effect that, in industrialized countries, the average age of first childbirth is increasing, and more women are having no children at all. That's one part of the story. The other part of it comes down to the limits of the human body. Because, of course, the bane of any woman who delays having children will be her age. Seven out of ten women surveyed who said they wanted children are concerned or very concerned about whether they will actually be able to have a baby by the time they get around to trying.

◉◉◉

For many animals, fertility fades alongside all the other functions of the body – a slow, steady decline that comes with age. A man becomes increasingly infertile with age, but may still be able to eke out a sperm and fertilize an egg into his seventies or eighties – around the time when the rest of his body is starting to shut down. A woman, as we have seen, loses the ability to reproduce some thirty years before then, and the process comes as something of a shock. Usually, all of a woman's other organs still function, her faculties are undiminished, her health remains robust – all the biological stuff remains more or less the same, except she can no longer have a baby.

Becoming sterile with age is not exclusive to humans, technically. Rodents, whales, dogs, rabbits, elephants, and domestic livestock experience an abrupt end to fertility too. Other primates, such as chimpanzees, gorillas, baboons, and macaques, also have a 'biological clock' and experience drops in fertility as they age. Still, female chimpanzees in the wild have been known to give birth at very advanced ages – even into their fifties and sixties – surpassing the UK natural-birth record set by Kathleen Campbell at the age of fifty-five. (Interestingly, the male chimps seem to prefer mating with these older females – a point that doesn't often get trotted out in evolutionary psychology. Perhaps, thinking like a Darwinist, it's because the older female chimps have displayed their fertility – most, it is presumed, have previously given birth.)

What is unique to humans among the primates is that our females have the potential to survive for a very long time after they become sterile. Though today humans enjoy longer and generally healthier lives – with access to better medical care than our ancestors had just three generations ago – the timing of the menopause has remained more or less the same since the days of the hunter-gatherers. Women stop being fertile around the age of fifty, and women live on average to be seventy years old – and much older in some countries. (In Japan, the average life expectancy for a woman born today is eighty-six years, and both the UK and the US come in at over eighty.) In reproductive terms, this is a waste of about one third of one's life. It's even worse when you consider the maximum life span for humans of about 122 years. If a woman lived that long, she would spend nearly sixty-nine percent of her life without the capacity to have babies. This isn't fantastical speculation; by 2025 the global population of women aged sixty-five or older is projected to be eight hundred and twenty-five million.

In theory, this long stretch of sterility in life may be an evolutionary adaptation, because women stand to gain greater genetic benefits in the long term by helping their children to reproduce successfully rather than by continuing to have more babies of their own, an idea called the grandmother hypothesis, first put forward by University of Utah anthropologist Kristen Hawkes. When an older woman becomes infertile, it helps to ensure her longevity – giving birth to babies with big heads via a narrow birth canal is a difficult endeavour at any age, but in older age the risk of death or medical complications is much higher. But the menopause also has a social side benefit: it frees up older women to care for their daughters' babies rather than compete with younger, more fertile women for sperm. (Of course, this assumes that the menopause marks a change in sexual activity, not just in reproductive capability. And the changes in hormones at this time do, indeed, make a difference.)

Compared to other animals, human babies are markedly dependent on adults (usually, mothers) for an extended period of time, which means that if a mother were to die, having a grandmother around may be key to a child's survival. Studies of pre-modern hunter-gatherer groups with no access to modern medicine have found that women with a prolonged, post-reproductive lifespan have more grandchildren, and that these children are significantly more likely to survive to adulthood if they have a grandmother's assistance. No other relatives have a similar effect.

For most mammals, a mother is more important than a father in determining an infant's survival into adulthood. *Homo sapiens* are considered to be unique, however, in the extent to which the family has traditionally provided help and care for our young. So opportunities for inter-generational co-operation may have been one of the evolutionary architects of women's early and prolonged period of sterility. A woman's reproductive success depended on having a sterile mother by the time she was caring for a brood of her own. The knock-on effects of menopause would have been very positive for a post-reproductive woman (provided, of course, she happened to have had a daughter in the first place). This may be why the menopause happens bang in the midlife of a woman, but nothing similar happens in a man.

The experience of going through the menopause is not simply a question of turning off the hormones that allow egg release and pregnancy; it is essentially a case of organ failure, with the ovaries shutting down completely, and some doctors consider the process to be as much of a medical threat as the failure of another organ, say, the gall bladder or kidney. The hormonal changes affect a woman's entire physical and emotional well-being. Post-menopausal women are at increased risk for several major diseases, including cardiovascular disease, breast cancer, and osteoporosis. The female biological clock counts down not just to the end of fertility but to the beginning of a new body, with different needs.

Still, although the onset of menopause seems to be an immutable part of nature, it appears to be amenable to manipulation. The incessant ticking can be speeded up or slowed down. We know, for example, that exposure to certain chemicals or radiation from cancer therapies can trigger premature menopause in young women. The menopause also happens earlier in women who smoke cigarettes.

Why, in evolutionary terms, the end of fertility in otherwise healthy women is so clearly marked has long been a great puzzle. A simple solution would be to work within the limits of what human biology allows. Fertility is subject to hormonal regulation, just like it is in every other mammal, so why not just whip up some hormones in the lab and trick the body into reproduction well beyond its 'sell-by' date? The production of hormones can be easily influenced by environmental conditions too.

For humans, though, reproduction is also socially regulated, from the acceptability of walking marriages among the Mosuo to the top-down condemnation of having 'unprotected sex' with someone who isn't a long-term partner. And social regulations don't live in some world separate from our biology. Take, for instance, the way in which sexually transmitted diseases affect fertility. STDs, such as chlamydia, cause infertility – and these diseases are on the rise. Chlamydia has long been known to cause damage to the Fallopian tubes, making women less fertile; recently, it has been discovered that the bacteria hurt sperm too – men aren't simply carriers of the disease. When men are infected, the bacteria physically 'sit' on sperm, gripping on to the sperm's tail at intervals, which reduces a sperm's ability to swim to an egg. The bacteria can also trigger sperm death. The jury is still out on whether treating chlamydia with antibiotics makes any difference to sperm health, as men who have had chlamydia can remain infertile long after the bacteria have been cleared from the reproductive system.

These are the easy obstacles. Humans may have a bigger problem.

A woman, even at peak fertility, needs sperm in order to make a baby. But fertile *Homo sapiens* males may not be with us indefinitely. The male Y chromosome appears to be hurtling down the evolutionary road towards what, some scientists predict, will be extinction. Our age as a species may be a factor in the future of reproduction as well.

The loss of the Y chromosome would certainly put humans in a complicated position. Over the past three decades, since the birth of Louise Joy Brown, technologies have emerged from the realm of science fiction into the reality of our hospitals and homes to make reproduction possible where it had previously not been. But can science get round the problems that come with getting older?

◎◎◎

Age gradually yet surely strips women of the key factors in being fertile – for instance, having good quality eggs, ovaries that function so that ovulation happens frequently (and efficiently), and a healthy womb. Medically speaking, thirty-five is the age at which a woman is branded as an 'older mother' in maternity wards, where even conventional childbirth becomes necessarily more medicalized. Decline seems to begin from the age of thirty, becomes more obvious between ages thirty-five and forty and increases quite dramatically after that. After thirty-five years of age, even if an egg does become fertilized, the ability of the embryo to implant in the womb decreases by around three percent each year. From the mid-thirties onwards, women are faced with six threats: declining fertility, miscarriage, genetic defects that accumulate with age, high blood pressure (which if unmanaged is very dangerous), stillbirth, and, rarely, death of the mother. Forty-one doesn't sound old, but it officially marks the point at which fertility stops and sterility begins. By forty-five years of age, there are only one hundred pregnancies for every thousand women having unprotected sex. Even IVF does not escape the age-effect. While a thirty-year-old will have about a thirty percent chance of becoming pregnant with this method, a forty-four-year-old faces a sliver of a chance: about 0.8 percent.

Yet, within the space of the past few decades, more and more women have chosen to have their first child later in life. In 1970, 11,704 American women had their first child between the ages of thirty-five and thirty-nine; by 1986 that number had jumped to 44,427, and by 1997 to 88,501. Twenty percent of women give birth to a first child when they are over the age of thirty-five. The birth rate for women aged forty to forty-five has risen thirty-two percent since 1999, and for women aged forty-five to forty-nine it has more than doubled. In 2008, seventy-one British women aged fifty or older gave birth for the first time. Many of these women were only able to become pregnant with medical intervention.

A similar story has, of course, unfolded in the US. The

number of childbirths among women over forty almost trebled between 1989 and 2009, from 9336 to 26,976, and the numbers of first births is even more staggering. In 1970, little more than 2400 women had their first-born child between the ages of forty and forty-four. By 1986, the figure had risen to 4419. In 1997, more than 15,550 women in this cohort gave birth to a first child.

In social terms, having babies later in life is no bad thing. Being able to choose when we have children is now possible because we have access to effective contraceptives, and about half of the rise in the age at first birth is attributed to a rising level of education. Women with higher levels of education are more likely to postpone having a first child, and so as more women have finished university or gone on to pursue postgraduate education, or have attained economic independence, the age of first childbirth has gone up. In 2006, fifty-eight percent of all UK higher education qualifications were awarded to women – just thirty years after the fight to gain admission of women to male-only colleges still lingering at Oxford and Cambridge. It may come as no surprise, however, that if you look at UK birth data, the map of births late in life follows the contours of social class. There is a higher average rate of forty-something mothers in the wealthier south of England compared to the poorer north, a pattern that can even be seen on a city level, between inner London and the relatively cheaper outer boroughs. Late motherhood is as much a marker of the better-off middle classes as a designer handbag. Infertility treatments do not come cheap.

◉◉◉

In order to allow women to have children late in life, we have to use medical intervention. That's because, any time between the ages of forty and sixty, most usually around age fifty-two, the majority of women will stop menstruating. After this happens, becoming pregnant naturally is extremely rare – close to zero percent after age forty-five, even if a woman pumps herself full of hormones to keep her ovaries functioning as they did earlier in life.

As we have seen, a female embryo's ovaries contain all of the eggs a woman will have in her life. These eggs, immature as they are at this stage, number between four million and seven million. But by the time a bouncing baby girl is born, she will already only have half as many. During puberty, around age thirteen, she will have on average 400,000 eggs remaining to cover her lifetime of fertility. If you do the maths quickly, you'll realize that should be enough to have one egg every month for thirty thousand years. As it happens, only four hundred to five hundred of this multitude of eggs are released in cycles of ovulation; the rest are unrelentingly destroyed – either literally imploding or dying from neglect if they aren't in the right chemical environment. The majority of eggs never actually mature to the point where they can be fertilized.

Starting from the age of thirty-five, a woman's eggs start committing suicide at an accelerated speed. So while at the age of thirty-eight a woman may have around twenty-five thousand eggs, by the time she is forty-five she will have closer to five thousand; by her early fifties, she will have only a few hundred left. So regardless of whether a woman becomes pregnant or uses an oral contraceptive, such as the pill, that stops her eggs from being released, a woman's supply of eggs is doomed to extinction by then. Importantly, those eggs that linger in the ovaries, taking their time to die, will stop doing what they are supposed to.

By and large, as we get older, the machinery in most of our cells simply doesn't work as well; eggs are not the only cells in a woman's body that go wrong with age, but they tend to be particularly affected by the process. This is because of the way in which egg cells develop. Unlike other cells, some of the eggs in a female foetus may have to wait for fifty years before they are triggered to mature, in readiness for fertilization. While they wait, the chromosomes they contain are lined up in a relatively orderly fashion on what is called a *spindle*, a structure that forms when a cell is dividing to create two (or four) new cells. The spindle helps guide chromosomes into newly created cells, so that their distribution is equal. Of course, having too many or too few chromosomes can be disastrous for health, so eggs and sperm must only ever contain twenty-three chromosomes or face serious repercussions. While the egg's chromosomes sit on this spindle, waiting for half a century for their chance, the DNA they carry may well be degrading. The spindle itself may also become damaged, meaning that the chromosomes are not divvied up as neatly as they should be when the egg finally divides. This is partly because, with age, there is also a decline in the levels of certain proteins, called *cohesins*, that normally hold chromosomes together by entrapping them in a ring – something that's essential for chromosomes to split evenly when a cell divides. This is one of the reasons why older women are more likely to produce abnormal eggs, which increase the risk of infertility, miscarriage, and birth defects, including the chances of having a baby with Down syndrome.

Down syndrome babies have three full or partial copies (called a *trisomy)* of chromosome 21, rather than the usual two. Though approximately twenty-five percent of all spontaneous abortions in the first trimester carry chromosome 21 trisomies, the chromosome error alone obviously does not terminate the pregnancy. Indeed, one in every seven hundred babies is born with chromosome 21 trisomy, and it remains the leading cause of learning difficulties and developmental delays in humans. Women with Down syndrome are sometimes able to reproduce. Most men with chromosome 21 trisomy are sterile from birth. Although the exact causes are not known, this infertility may be caused by hormonal deficits, changes to the shape of the gonads, or problems generating sperm.

Other chromosome trisomies are likely to have devastating effects too. For instance, embryos that have one copy or three copies of chromosomes 1 or 19 end up being miscarried before a woman even thinks to perform a pregnancy test. Similarly, in nearly all recorded cases of girls who carry XXX instead of XX in their sex chromosome, the extra X has come from the mother, not the father, and those mothers were usually older than average. Most XXX girls are of normal weight, height, mental function, and fertility, but they tend to experience a very early menopause, around the age of thirty. Somewhat like taking a too-high dose of a medicine, an extra chromosome translates into a too-high dose of certain genes, and abnormally high dosages are ultimately detrimental to a child's health. In this case, the more X chromosomes a girl carries, the more severe her symptoms will be. That's not to say that sperm cannot carry this chromosomal corruption, however. There is a significant increase in rogue X chromosomes found in the sperm of older men. A son who carries an extra X chromosome will suffer from Klinefelter syndrome, and will likely have difficulty producing sperm, to some extent; generally, he will be sterile. The condition affects one in a thousand men.

The eggs may be damaged by the environment of the ovaries themselves, which are no longer a safe haven as a woman ages. For example, oxygen levels, pH balance (whether acidic or basic), and hormone concentrations are in flux, and each of these in turn can make it more difficult for the eggs to separate the chromosomes from each other normally in pairs. If the chromosomes stay together rather than splitting apart, then one of the two new cells will have two of the same chromosomes and the other will have none. As the number of eggs ripening in each cycle drops in the run-up to the menopause, there is an increase in chromosomal abnormalities within the eggs. In fact, in IVF programmes the vast majority of eggs that come from women who are thirty-seven years or older have too many or too few chromosomes, as well as mutations in their DNA and in the machinery that controls the way in which this DNA is expressed. For example, ageing eggs are more prone to producing *hydatidiform moles*, the grossly distorted embryos that result when genes that are normally silent and locked become active, even without the proper imprinting that tells the DNA what to do. When this happens, the renegade DNA behaves as if it has come from the father rather than the mother, so that the embryo is effectively working with two sets of paternal genes. All that can come out of this combination is a mass of tissue inside the womb that, as of now, can never develop into a baby.

An ageing woman's eggs no longer mature in readiness for fertilization, as younger, healthy eggs normally would, when they are exposed to follicle-stimulating hormone, or FSH. As the menopause approaches, the ovaries stop responding to FSH (they also stop producing oestrogen and progesterone, as we saw earlier). In response, the body produces more and more FSH – ticking and tocking louder and louder. The interactions between the hypothalamus, the pituitary gland, and the ovary change as well. The seamless orchestration of hormones necessary for successful fertilization and pregnancy transforms into a cacophony. But it is the eggs and their corrupt chromosomes that are most at fault. When an older woman is implanted with young eggs – even just the cytoplasm, that cellular soup inside an egg minus its DNA – they become pregnant, despite all of the other changes going on in the body. In fact, older women actually stand a better chance of becoming pregnant with donated young eggs than younger women do of conceiving naturally.

Something doesn't seem quite fair about these facts of fertile life, when you start to think about them.

⊙⊙⊙

It has long been reasonably obvious that for women, age is the main factor in the loss of fertility. But matters are less clear when it comes to men. In contrast to women's reproductive ability, male functions do not cease so abruptly – there is no single event that parallels the menopause. This means that throughout their lives, men continue to produce sex hormones and to generate sperm. And yet, the effects of paternal age on a couple's fertility are significant.

A healthy couple in their mid-twenties has only a twenty to twenty-five percent chance of establishing a 'natural' pregnancy in a given month, while a couple aged forty can only match that chance through the use of IVF and other techniques. Because older women tend to have male partners who are around the same age as them, or older, older couples also carry the added risk from the greater number of genetic mutations that occur in the sperm of older men – the curse of corrupt chromosomes.

It is now understood that having a male partner over forty years old is an important factor in failing to conceive, and when a man is over fifty, there is also a significant decrease in how many embryos form properly and how many babies are born alive. If a woman's male partner is over thirty-five, there is a higher risk of seeing the pregnancy end in miscarriage than for those whose partners are younger, regardless of the woman's age. If the pregnancy succeeds, sperm-based mutations may lead to a range of lifelong conditions. Autosomal dominant diseases, such as short-limbed dwarfism (achondroplasia) and

Marfan syndrome, affect connective tissue and can cause problems in the skeleton, eyes, heart, blood vessels, nervous system, skin, and lungs. Some genetic disorders that arise as fathers get older behave a bit like a jammed CD, with sequences of the DNA code repeating when they shouldn't. These conditions include Fragile X syndrome, the most common cause of inherited mental impairment; myotonic dystrophy, a disorder of the muscles and other body systems; and Huntington's disease, a currently incurable condition that causes deterioration and gradual loss of function in the brain.

When it comes to making babies, time is not on our side, whether you are dealing with egg or sperm. While the link between age and infertility is certainly biological, some people are infertile early or throughout life. Leaving the decision to have a baby until the thirties or forties means that the underlying cause of a person's infertility won't be identified until it may be too late to identify or correct. That is to say, most people won't find out they're infertile until they start trying, whenever that is. Around one in six couples who cannot establish a pregnancy on their own will seek medical assistance, and the average age for receiving IVF procedures in the UK is thirty-five. Sometimes the problem is a minor one, but that is not always the case. For around one quarter of people who find they cannot get pregnant, the problem cannot be pinpointed at all – the dreaded 'no diagnosis'. Slowly but surely, scientists are uncovering what may be behind these mysteries.

◉◉◉

There is a growing list of genes that have been found to be key players in regulating when a woman is fertile and when she is sterile.

A gene on the X chromosome, *FMR1*, for example, is coming into use as part of a genetic test that aims to predict the rate at which a woman's egg supply is running out. *FMR1* is known to help regulate the transition of eggs from immaturity to maturity. The sequence of chemicals that spell out the *FMR1* gene contains repetitions, and women with a version of the gene that contains more than two hundred repeats of the DNA sequence CGG are likely to have Fragile X syndrome. But there are also women who have fifty-five to two hundred CGG repeats – not quite enough to disturb the gene and cause mental impairment, but enough to put the carrier at increased risk of experiencing an early menopause. If a woman has between twenty-eight and thirty-three repetitions this leads to abnormal levels of anti-Müllerian hormone, or AMH, which fluctuates throughout life. Healthy women with low levels of AMH for their age seem to hit menopause earlier; they also have fewer eggs, lower fertilization rates (whether through 'natural' means or IVF assistance), generate fewer embryos, and have a higher incidence of miscarriage during IVF transfers. Women with insufficient AMH have half the number of successful pregnancies compared with women with high AMH levels. In fact, AMH levels are usually a better tip-off than a woman's age in guessing how successful IVF will be, ranging from how many eggs will be harvested from her ovaries to whether she may miscarry once an embryo is transferred from a Petri dish to the womb.

But *FMR1* is not the only genetic clue to a woman's fertility. It appears the *BRCA1* gene mutation – widely known for its role in breast cancer and ovarian cancer – may offer information about the risk of premature ovarian ageing. Normally, *BRCA1* rallies other genes in the cell to repair damaged DNA. When the *BRCA1* gene itself is damaged, or when damage accumulates on its chromosomes, you start to see the growth of abnormalities and, later, tumours. An inability to repair

DNA seems to be an important part of why women's ovaries age, and that's where *BRCA1* comes in. Manipulating the genes that are involved in DNA repair could be one way to avert failing ovaries in the future. Further, new bits of genes on chromosomes 13, 19, and 20 have been found that influence the age at which a woman experiences menopause, as well as ageing-related diseases such as breast cancer, osteoporosis, and cardiovascular disease.

For a woman facing the dilemma of focusing on career or childbirth in her twenties or thirties, being able to predict the age at which she might begin to have serious difficulty in becoming pregnant would be the holy grail of family planning. The decision isn't binary, however: it's not a choice between pregnancy now or never. A woman who wanted to focus on her professional ambitions for the next decade already has the option of freezing her 'young' eggs for use in the future. But at a cost of £3000 per attempt, and some women having to undergo three rounds of extraction to get a good harvest of eggs, most women, without sure information, will be likely to defer until later. In the future, gene therapy may also be developed to reinstate the functioning of what is normally lost after menopause, affecting the treatment of fertility and age-related diseases. Such therapies might even extend the life span of a woman's ovaries and allow a woman to remain 'naturally' fertile for far longer than has ever been possible. An 'old' mother in the future will probably bear little resemblance to the 'old' mothers who hit the front pages today: she would not have needed IVF or a donor egg, because she will be able to use her own, without detrimental effect on her health or life expectancy. Scientists are still uncovering exactly which genes will be useful or amenable to manipulation, but the research is already in full swing.

◉◉◉

If the human X chromosome sometimes harbours genes that can cause problems for female fertility, the Y chromosome can be viewed as a disaster zone. The Y chromosome, which once contained as many genes as the X chromosome, has deteriorated so much over time that it now contains fewer than eighty functional genes compared to its partner, which is large and packed with more than one thousand. This deterioration, according to geneticists and evolutionary biologists, is due to accumulated mutations, deletions, and anomalies that get stuck, in a way: they have nowhere to go, because the Y chromosome doesn't swap genes with the X chromosome like every other chromosomal pair in our cells do.

With such a small number of genes to carry, the Y chromosome is small. It's also peculiar, filled with many repetitive DNA sequences but not many genes. Unlike the X chromosome, whose genes display a variety of general and specialized functions, the Y carries the codes for only forty-five unique proteins. These proteins are the blueprints essential for the male reproductive system, particularly those important in sperm development. Of course, when an egg is fertilized, the sex chromosomes are matched up. In a woman, the two XXs pair up easily. But because the Y and the X are so different, in size and information, the match isn't quite right. So in a man, the X and the Y align only in a small region, where you could say the chromosomes are singing from the same hymn sheet. This has helped to perpetuate the diminution of the Y – the parts of the chromosome that don't match up with the X aren't necessary; they're expendable. Thus, the Y has slowly but surely become smaller and less genetically rich compared with its sex partner.

Technically, it's not simply having a Y chromosome that makes a person male, it's having the right bits of the Y – the right key genes, known as testes-determining factors. The most important gene of this group is *SRY*, for sex-determining region of Y, but there are almost certainly other genes that scientists have yet to identify. One case dramatically illustrates the role of *SRY*: the rare sex chromosome disorder known as de la Chapelle syndrome, also called XX male syndrome. Individuals with the syndrome appear to be male, though they have two X chromosomes and no Y – just like a woman. The critical difference is that the *SRY* portion of a Y chromosome has usually become attached to one of the X chromosomes. That small error effectively converts an X chromosome into a Y, female into male. The genetic mutation is sometimes evident in a short stature, an abnormally shaped penis, and the appearance of breasts. Curiously, there are a considerable number of cases in which XX males do not carry the *SRY* gene. Instead, these men have a closely related gene, called *SOX9*, which also causes skeletal deformations. As a result of these genetic anomalies, somewhere between one in nine thousand and one in twenty thousand men have XX chromosomes, rather than XY. Maleness – to some extent, depending on how you define it – is perfectly possible without a Y chromosome, or even the *SRY* gene.

Of course, one way in which we define maleness, socially, involves sexual reproduction. People with XX male syndrome have little or no detectable sperm in their semen, making them effectively sterile. Not having a Y chromosome is bad news for fertility. Take XX male cocker spaniels, for example – without the *SRY* gene, many turn out to be infertile hermaphrodites. There have also been several reported cases of XX male farm animals, including among pigs and goats. This is unfortunate news for the breeders who own a particular animal, but is not yet a threat to their livelihoods – although that may not be true in the future.

Still, against the odds, the human Y chromosome does not seem to be taking its destruction lying down. Even though it is at a distinct disadvantage by not having a perfect partner match in the cell, it has developed ways to ensure its survival. To get rid of accumulating damage and mutations, the Y chromosome has been swapping its bad bits with intact bits of itself. While this has fixed the problem to a certain degree, it has introduced other places where mistakes can happen. Once the DNA on a Y chromosome is broken up so that it can be swapped, the pieces sometimes end up being put back together incorrectly. This self-protection stratagem triggers a whole range of sexual disorders in otherwise healthy men, including less sperm production, sterility, and sex reversal, as in de la Chapelle syndrome.

Should the Y chromosome crumble into scant genetic bits and bobs, and then disappear altogether, it would not necessarily mean the end of a species. There is hope. Some animals, such as the mole vole (*Ellobius lutescens*) and the Japanese Ryukyu spiny rat (*Tokudaia osimensis*), have two sexes but no visible sex chromosomes. The voles, a rodent that burrows underground throughout a wide swathe around the Caucasus mountains, are extremely interesting for one reason: it is impossible to distinguish a female from a male by looking at their chromosomes; both carry only a single X. What is more, scientists have been unable to pick up any bits of *SRY* gene on any chromosome in the mole vole species. Since *SRY* normally acts as the primary switch that initiates the development of a testis out of the undecided mass of cells in an embryo, you would expect to see it in an animal that looks male. It's not there.

Unlike with the XX male cocker spaniels, pigs, and goats, sex reversal is the norm of the male mole vole, not the rare exception. How does the species survive? Male mole voles do have small testes and problems in generating healthy sperm, but no hermaphrodites have ever been recorded in these animals. It appears, however, that testosterone is not very efficient in the males; the prostate gland, in fact, seems to be insensitive to the effects of the hormone. Still, in captivity at least, mole voles seem to have little problem breeding. Fifty percent of early embryos perish, but certain mole vole couples have been seen to give birth to a litter of up to six pups – and they did that every four to six weeks, eleven times.

The mole vole story may give us a hint about the future of humanity. Even if the human Y were to disappear, it wouldn't necessarily follow that men will vanish, too, though male fertility would likely become substandard, adding another spanner in the works of making babies whenever you like. Infertility might one day affect all young, otherwise healthy men. And if the trend for women having babies later and later in life continues, the health and quantity of their eggs will also be an issue. The limits of time is undeniably something we need to address, perhaps by correcting genetic mistakes in embryos, a technique that is already being discussed but raises the spectre of eugenics in many circles.

Better answers will probably come from social policy rather than biology: discussing with young people the biologically optimal time to have babies, at the same time as they are taught how to prevent pregnancies; more support for people who have children at a young age; more extensive childcare, benefits, and incentives to allow for family and work to exist side by side. After all, right now, IVF treatments are invasive, difficult, and very expensive. For many people, however, 'losing' their youth to parenthood is neither an ideal nor a practical life choice. It's very hard to get away from the uncomfortable fact that it is educated women who tend to have babies later in life. It has even been suggested that an effective way to control overpopulation would be to increase women's rates of literacy. In any case, many women who have access to education and work opportunities are not going to turn them down in order to have babies at a young age and at the risk of slipping behind their male colleagues on the career ladder. The goal should be to give everyone these opportunities, not to snatch them away from those who have won them.

In a world where men and women now very often face the same social prospects, our reproductive biology has not kept pace. Normally, men produce sperm, women under thirty-five have eggs, women bear children, and men cannot. There is a clear division between the sexes. That, however, is set to change.

8

REAL MEN BEAR CHILDREN

Women's liberation is just a lot of foolishness. It's the men who are discriminated against. They can't bear children. And no one's likely to do anything about that.

Golda Meir, quoted in *Newsweek*, October 1972

In 2008, scientists at the New South Wales Department of Primary Industries, in Australia, developed the first artificial womb. It was a plastic container specially designed to hold fluids, bacteria, and the other stuff that is needed to mimic the conditions found inside the mother as an embryo develops. It was a phenomenal breakthrough. Especially if you were a grey nurse shark, the species for which the womb had been developed.

For aquatic ecosystems scientist Nick Otway, the artificial womb was a tool for addressing a terrible problem. The grey nurse shark, also known as the sand tiger shark, has roamed the world's oceans for more than seventy million years, but in the past century, the species has been decimated by increased fishing. Though it is not the target of the fishing itself, the animals get tangled in commercial fishing nets or are mistakenly caught on hooks by recreational sportsmen. As a result, the shark is now listed by the World Conservation Union as globally vulnerable and as critically endangered in eastern Australian waters; its risk of extinction is high.

Part of the problem is the way in which grey nurse sharks reproduce in the wild. When a female becomes pregnant, dozens of embryos are produced. But at the end of a gestation of nine to twelve months, the lengthy labour expels only two pups, each about one metre (just over three feet) long. The mother then enters a year-long rest before the next pregnancy. This means that the shark's birth rate is low, and births are few and far between. And it is lucky that the female produces two pups at all, and not just one – because the female grey nurse shark has not one but a pair of separate wombs, so these offspring develop in isolation. And there is a reason for that.

The grey nurse shark is what most people think of when the word 'shark' is mentioned, despite the cinematic fame of the great white. In fact, historically, most sharks were labelled the grey nurse – especially if they had taken a bite out of a person. It's easy to see why – their menacing appearance is compounded by the fact that this is one of the few shark species to display its impressive jaw of jagged teeth all the time. But most attacks attributed to the grey nurse are now considered to have been committed by other animals, and it is believed that they have never actually been involved in a case of an unprovoked attack on a human. But while their savage reputation is unwarranted when it comes to humans, you can see how it might have been earned when you consider exactly how grey nurse foetuses develop in their mothers' wombs – through cannibalism.

Until they reach a length of about six centimetres (just over two inches), the embryonic grey nurses are nourished by the egg's store of yolk. When they hatch from the egg, the foetal pups swim into the so-called nurseries inside the mother's body – what we would call the wombs – but very little nourishment is left there. Luckily, by the time the foetuses grow to about ten centimetres, they have developed a nice set of menacing jaws. To feed themselves, the foetuses begin to eat the remaining egg capsules, containing eggs and younger embryos, and then, once they have consumed all of that, they attack the other foetuses in the womb – their siblings. The first foetuses to hatch will be the largest of the batch, so the baby shark that manages to eat all of its siblings is likely to be the one that was fortunate enough to develop first. This cannibalism does mean, though, that out of the up to eighty embryos at the beginning of the pregnancy, only one victorious pup is left in each womb at the end. As the other foetuses aren't enough to feed the pup through to delivery, the mother shark nourishes the two remaining pups, safely separated from each other's jaws in their separate wombs, with a continuous supply of freshly produced eggs. Every grey nurse shark has survived at the expense of dozens of its siblings – and thus no one female can produce more than two pups every two years.

These are not figures that can sustain an endangered species, and that's how the idea of an artificial womb came about. The idea was suggested, in general terms, by an Australian government minister in charge of fisheries but who was himself a farmer. As such, he was well versed in the manipulation of reproductive strategies, such as IVF, as a way of addressing breeding problems in cattle and other livestock. He challenged scientists to create a similar intervention that might increase grey nurse shark numbers. Nick Otway answered the challenge by looking for a means of pulling those dozens and dozens of shark embryos out of the mother and giving them a fair chance of coming to full term, away from the jaws of their siblings.

To create a successful surrogate for a shark womb, Otway, his research partner Megan Ellis, and their team first needed to figure out what a shark womb is like. What is the chemical composition of the fluids in the womb, and of the eggs that the mother feeds the foetuses? What amount of oxygen exists in the womb, and what is the fluid's temperature? Are there types of bacteria that should be present, because they exist in the womb's natural environment and might play some crucial role? What is the consistency of the walls of the womb, and does the mother's body allow extra nutrients to be supplied through it? Do any or all of these factors change at different stages in the course of a pregnancy? And could scientists invent an artificial fluid to match the womb fluid, or re-create the overall environment?

To test the artificial womb prototype they developed, the scientists turned not to the endangered grey nurse, but instead to the related wobbegong, or carpet shark. The wobbegong is more docile in outward appearance than its cousin: flat, sand-coloured, and patterned, with short catfish-like tentacles surrounding its mouth; it keeps its sharp teeth hidden from view. It's also smaller and easier to handle than the grey nurse. Internally, the wobbegong is simpler in structure, but there are similarities between the two species. Of special note to the research team, the wobbegong reproduces more frequently than the grey nurse and was in no danger of dying out.

In a surgical procedure, Otway and the team removed six embryos from a wobbegong and placed them in their specially designed tank filled with some artificial womb fluid, some bacteria, and other elements. After a normal period of gestation, the pups were 'born' through a tube that connected the grey box with another one containing ocean water, similar to that which would be found where naturally developed pups would be born. The pups were reluctant to leave this rudimentary 'womb', even trying to swim back into the womb tank after making their initial exit, but they eventually entered their new tank before being transferred into a more natural habitat. Proud as any new father, Otway said he was relieved, pleased, and even amazed that everything had worked.

Next, Otway would focus on removing the embryos from the mother's womb earlier and earlier – a move that would add layers of surgical complexity to the process, since the scientists need to ensure that the delicate external yolk sac, as an essential source of nutrition, remains connected to the embryos as they are removed. Once this is accomplished, the embryos will also have to be tethered to the artificial womb in some way, to stop them from detaching themselves and swimming away from their nourishment. If possible, Otway hopes to extract the embryos so early that one day they might gestate completely in the artificial womb.

Otway's artificial womb may be a novel idea for shark conservation, but bypassing a woman's body is no new ambition when it comes to human reproduction. In 1924, evolutionary biologist J. B. S. Haldane coined the term *ectogenesis* to describe how pregnancy in humans could be provided through an artificial womb. In a fictional account, he had two future scientists describe the birth of the world's first ectogenic child. 'Now that the technique is fully developed, we can take an ovary from a woman, and keep it growing in a suitable fluid for as long as twenty years,' one of the characters announced. This, by the character's calculations, would result in 'a fresh ovum each month, of which 90 percent can be fertilized, and the embryos grown successfully for nine months', at which point they could be 'brought out into the air'. Haldane imagined that artificial wombs might become so popular by 2074 that only a small minority, 'less than 30 percent of children', would then 'be born of woman'.

Otway and Ellis finally reported the successful 'artificial' birth of their sharks late in 2011. But they had started the project early in 2008, and it had been a turbulent process, with design failures along the way. Many embryos had perished. If creating such a device for a shark has been challenging, could an artificial womb be viable for humans?

In some ways, the female grey nurse shark's reproductive system is similar to a human's, as eggs are produced in ovaries and pass down tubes towards the womb (whichever womb that may be). As we saw, the pups first develop while still inside these eggs, surviving on the egg yolk, before the cycle of cannibalism begins.

But, in most ways, the grey nurse shark nurtures its foetuses in a far less complicated manner than humans; for whom getting nutrition is not as simple as ingesting egg yolk (or siblings). For a start, the shark does not have a placenta – the complex, specialized organ that is created from the fertilized mammal's egg in order to sustain the embryo. Because the placenta is made from the fertilized egg, it contains both the mother's and the father's DNA. Very early in the pregnancy, the placenta sends out a system of blood vessels to penetrate and dock into the mother's womb, through the umbilical cord. The foetus acquires oxygen and food – gets rid of its wastes – through the placenta, which should provide life support for forty weeks, until the foetus grows and develops to a stage where it is able to perform these functions on its own. A human artificial womb would need to replicate all of the placenta's functions, not just the womb's fluids, bacteria, and other stuff essential to the making of life.

Still, babies born prematurely have, for more than a century, been reared through several weeks of life despite being unable to live completely on their own. In 1975, Kim Bland became the first child to survive after being born just six months into pregnancy. He weighed little more than a couple of hundred paperclips – and had he been born ten years earlier, there would have been no medical efforts to keep him alive, because at that time, babies who weighed less than a kilogram (thirty-five ounces) were not considered to be viable outside of the womb. And on 24 October 2006, Amillia Sonja Taylor was born after less than twenty-two weeks in the womb, the youngest birth yet to survive. Born in Miami, Florida, where the law does not support the revival of a foetus less than twenty-four weeks, she owed her resuscitation to some ambiguity around her size (she appeared to be a bit older in the sonograms), and to the fact that her parents, Sonja and Eddie, were liberal with the truth surrounding her conception (the hospital was under the impression that Sonja was twenty-one weeks pregnant when she arrived for emergency delivery). At birth, though, Amillia was clearly tiny: weighing less than three hundred grams (ten ounces) and measuring only twenty-four centimetres (9.5 inches) long. Her body was a snug fit in the palms of an adult's hands, and her feet were the size of one of the phalanges of an adult's finger.

Both Amillia Taylor and Kim Bland were only able to survive because they were nurtured in an artificial environment already within our reach – the incubator. Like the stories of an artificial womb, the technology and benefits of the earliest incubators were a genuine marvel to the public when they initially came into use. But then again, they were brought to the world's attention in a very bizarre manner.

◎◎◎

In 1897, a German paediatrician named Dr Martin Arthur Couney moved his showcase of medical specimens from Coney Island in New York City to the Victorian Era Exhibition, at Earls Court, London. Rather than the usual monkeys, midgets, minstrels, and Moors, visitors to Couney's were met with a true spectacle: a room neatly arranged full of large, glass-lidded wooden boxes, each containing a tiny baby, his 'child-hatchery'.

The boxes were based on the original designs of the French obstetrician Etienne Stéphane Tarnier, who had realized that keeping premature babies warm was not enough; they had to be provided with isolation, excellent hygiene, appropriate feeding, and a warm, humid atmosphere in order to survive. Tarnier had studied a warming chamber used for rearing poultry at the Paris zoo. In 1880, he built his first enclosed wooden box for infants, outfitted with a compartment to hold a hot-water bottle that could warm the space without letting in germs. This crude incubator reduced the mortality of premature babies by nearly one third. Thirteen years later, Tarnier's assistant, Pierre-Constant Budin, improved the basic contraption, adding a thermostat and natural-gas heating and more windows through which the babies could be observed – an innovation that his student, Martin Couney, must have applauded. Observed the babies certainly were – in Earls Court alone, the display drew crowds nearly four thousand strong.

The babies with whom Couney filled his incubators had been supplied by a Berlin hospital. As they were born prematurely, they were fully expected to die prematurely, too, which released the 'incubator-doctor' from liability for their deaths. Yet it was claimed that every one of the babies from his exhibitions had survived. With the money he made from his various circuses, Couney purchased more glass boxes for his hospital. His attempts to manufacture an artificial, independent environment for growing babies had proved a success.

A more technologically sophisticated means of sustaining the premature was developed in the late 1950s. This comprised a mass of machinery – plates and gaskets clamped together, with connectors for blood and gas; stainless steel plates and bolts; fixed volume gas exchangers; pressure transducers; and water baths. This incubator was used in experiments conducted on lamb, goat, and rabbit foetuses that were extracted very early from the mothers' wombs. The incubator was meant to replicate the idealized environment within a mother's body, and the age of foetuses for which this became possible was pushed further and further towards the beginning of life. The ultimate aim, of course, was to translate this technology into saving human babies' lives.

A baby who is born full term, after spending thirty-seven to forty weeks in a woman's body, should have lungs that are sufficiently developed to support breathing air by him- or herself. The lungs of babies born at around six months, however, are prone to collapsing between breaths. This problem can be overcome by providing the baby with a ventilator, which mechanically keeps the lungs slightly inflated between breaths, and by treating the lungs with a chemical called a pulmonary surfactant (which reduces the surface tension in the lungs) that would have been produced naturally, had the lungs been able to develop fully in the womb.

Underdeveloped lungs are a major battle front in sustaining the very premature; indeed, newborns' deaths because of respiratory failure have been recognized since ancient times. In the third millennium BCE, the legendary Chinese emperor and philosopher Huangdi reportedly noticed that this fatal syndrome occurred more often among infants born prematurely. Techniques for artificially reviving breathing in newborns date back to Soranus of Ephesus, who lived in the first century CE. Soranus even criticized 'the majority of barbarians' for the evidently common practice of immersing an infant in cold water to encourage them to breathe. And in the fourth century BCE, the father of medicine, Hippocrates, appears to have been the first to describe an intervention that is still in use to this day – inserting a tube into the trachea to support ventilation.

Even into the early eighteenth century, divine intervention

was mostly given the credit for successful resuscitation. From the mid-1600s, midwives were trained to use mouth-to-mouth resuscitation as an attempt to awaken stillborn infants – with little luck. The technique seemed so clearly destined for failure that the Royal Society, dedicated to the discussion and promotion of scientific topics, branded it nonsense, stating in no uncertain terms that 'life ends when breathing ceases'.

Others were more scientific in their approaches, though some of their solutions were often bizarre, and certainly amusing. In 1752, for example, the Scottish obstetrician William Smellie outlined the standard repertoire for treating apparently lifeless newborns: 'the head, temples and breast rubbed with spirits; garlic, onion or mustard applied to the mouth and nose'. (Smellie also advocated a form of artificial respiration, and also the application of a straight endotracheal tube for resuscitation, much as is still used today.) Doctors have pried bellows up the nostrils; wafted brandy mist under the nose; shaken the body or swung it upside down; rhythmically pulled the tongue in and out; tickled the chest; tickled the mouth; tickled the throat; yelled. They have also tried dilating the rectum using a raven's beak or a corncob, and blown tobacco smoke up the rectum with a clay pipe.

Fortunately, corncobs have fallen out of fashion, and artificial respiration using a ventilator – a mechanical device that fills the lungs with air – is now the accepted course. Unfortunately, the ventilator has not overcome one of the problems in resuscitating very premature babies: odds are that the life-saving device will irrevocably damage the delicate lungs, with serious side effects. If the lungs are damaged, that means less oxygen gets to the brain, increasing the likelihood of mental impairment.

So doctors and scientists started looking for a method that more closely simulates how air enters the foetus's lungs in a mother's womb. There, the foetus receives oxygen through the umbilical cord via the placenta, but the lungs are filled not with air but with fluid. This amniotic fluid is the 'water' that 'breaks' at the beginning of labour. At around six months, the foetus can use its own lungs to absorb oxygen, much as adults do, but it still continues to absorb oxygen from the amniotic fluid in which it lives.

Naturally, when scientists came up with a gentler alternative to forced ventilation, the method they chose involved delivering liquid oxygen directly into a premature baby's lungs. The result is the ECMO (extra-corporeal membrane oxygenation) machine, which is essentially an artificial lung. For the ECMO to work, surgeons must attach the machine's pump to the blood vessels in the baby's neck or groin. But this is not free of risk either. Using ECMO can trigger bleeding, blood clot formation, and infections, and lead to transfusion problems, so although the chances of survival are much higher, doctors are still searching for an even cleverer technology that can circumvent the threats that arise when a mother's womb is no longer the foetus's home.

Of course, the fragile state of the lungs is not the lone concern for the parent of a premature baby; the very thing that defines us as human – the brain – also needs a great deal of medical attention. It is only at thirty-seven to forty weeks of gestation – full term – that the brain passes certain key milestones that allow it to provide support for life outside the womb. These include greater myelination, where brain cells become coated with myelin, a substance that helps them to transmit signals faster and more efficiently – including processing sensory information and sending directions for responses, the sort of activity that permits us to pull our hand away from a flame, and thus survive threats to life. Because the brain develops over a more protracted timespan than the other organs, it is only in the third trimester of pregnancy, about twenty-eight to forty weeks after conception, that the striking growth and refinement of the brain's wiring takes place. And because the brain is relatively immature at birth, it is more susceptible to injury from premature arrival in the outside world.

Babies born before full term will fare very differently, depending on when they are born. At thirty-six weeks, a premature baby will probably be slow to feed. Before thirty-three weeks, however, a baby will need to negotiate more serious problems, including those immature lungs. And at twenty-eight weeks or younger, a baby will face some very significant problems – though he or she will still have a remarkable survival rate of up to eighty percent, thanks to modern medicine.

One of the most common issues among the very premature is bleeding in the brain. Although doctors do not know exactly why this happens, it increases the risk that a child born prematurely will sustain a cognitive or neuromotor disability, such as cerebral palsy (including the inability to walk), blindness, profound deafness or mental retardation. Forty-one percent will be diagnosed with such an impairment by the time they reach school age, meaning that within the current bounds of care, most will face a lifetime of disability. This is compared to only two percent of classmates born at full term.

Around five percent to nine percent of all babies in developing countries, and twelve percent of babies in the US, will be born prematurely. In some places, the increase in pre-term births is related to an increase in the number of multiple births of babies, often conceived through assisted reproductive technologies such as IVF and ICSI. But there are socio-economic factors involved too. For example, the average birth weight of a baby should be between three and four kilograms – from six pounds eight ounces to just shy of nine pounds. Though seven percent of babies in the UK weigh less than 2.5 kilograms, or five pounds eight ounces, at birth, this percentage rises to ten percent in deprived areas, such as Hackney, a poorer borough in east London.

In Hackney, I visited the intensive care unit of a neonatal ward at a large hospital. The ward contained several rooms for the care of the very premature, and all were furnished with incubators that contained babies of different ages. Above the clear plastic boxes (the material of choice, rather than glass and wood) were monitors, fitted with flashing lights and steady electronic beeps that announced every heartbeat, breath, and deviation in blood pressure. Every so often, a clutch of staff clad in blue gowns would wheel in a new incubator with an even newer baby, tubes threaded gingerly to the nose and veins to deliver essential oxygen and food. A new red-eyed parent or set of red-eyed parents would stand by, asking questions. A neonatal nurse would be assigned to stand sentry over the box – checking, measuring, tending. The nurse would work to make the incubator cosy, humid, and homely, like the environment the baby had left behind too soon. In the smallest of the ward's rooms, the babies were so young that many still looked like foetuses. One in particular was mesmerizing, a perfectly formed tiny doll with translucent skin, his delicate hands and miniature feet wriggling occasionally. It was only twenty-four weeks since he had been conceived. It felt voyeuristic – like looking inside his mother's womb, at a being no one should normally see for another three months. Intimate, remarkable, beautiful. And a revelation.

◎◎◎

For the survival of babies born prematurely, the incubator is a triumph of bioengineering. Yet what can be achieved with incubators is still very limited. One of the paediatric consultants on the Hackney ward described how, although modern incubators look sleek and efficient, the way they function and what they can provide has essentially not changed in decades. While the incubators can provide warmth and humidity, they still cannot give any of the nutrients necessary for growth. Instead, a premature baby must have all those tubes inserted into his or her body to deliver 'parenteral nutrition', that is, the complete nutrition provided intravenously, via needle-like catheters inserted directly into the veins and bypassing the digestive system altogether. Yet more tubes passed through the nose and threaded to the stomach will give the child milk, if tests show its digestive system is able to take it. This care is incredibly challenging due to the extreme immaturity of the baby's gut. In a mother's womb, the stomach and gut will not digest food at such an early stage of gestation, but once out of the womb, a small amount of the mother's breast milk will be used to acclimatize the baby's stomach to its new environment. The baby will also be sedated, at least some of the time, to stop him or her from pulling the tubes out, and to decrease or prevent any discomfort or pain. Moreover, infection around the tubes is a serious threat, and can lead to severe problems for the child's future health, should he or she survive. However you look at it, the incubators we have today are still a poor substitute for the relative security provided by the mother's womb.

Over those same decades, doctors have been attempting to secure the viability of ever-more premature babies. To succeed, they will have to invent an incubator that is more womblike – something almost like the womb itself. One unlikely source of data into how a womblike incubator might work has come from studies of miscarriages that have occurred at the very earliest stages of pregnancy. By looking at pregnancies that have failed, researchers have been better able to understand how embryos implant into the womb. Some day, with this information, embryos might not only be created in the laboratory, through in vitro fertilization, but even be attached to an artificial placenta in an artificial womb and gestate there until they are ready to be born. Still, it's one thing to cultivate sharks successfully in a man-made environment; it's another to nurture humans in one.

Scientists in Japan and the United States are experimenting to find out if an artificial womb for humans might be feasible, using cells both in Petri dishes and in living animals to copy the inner workings of mammals. With all mammals, it is vital that newborns are able to discard the placenta when they leave the natural womb. The placenta, after all, is the life-support mechanism that allows a fertilized egg to develop into an embryo; it is also the frontline of development in protecting the foetus from infection and in providing nourishment. For this reason, when in the 1960s researchers first began to toil towards creating a machine that would stand in for the essential placenta, they faced many of the problems that intensive care doctors face when human babies are born prematurely. These problems proved especially challenging with their chosen subjects – goat foetuses.

During a further series of experiments in the 1980s, a team led by the late Professor Yoshinori Kuwabara encountered several serious setbacks. In particular, the goat foetuses moved, as all babies in the womb do – and some quite vigorously. During incubation, and especially once their condition was stable, the goat foetuses showed a variety of movements that would have been absolutely normal behaviour had they been in their mothers' wombs. They rolled their eyes, moved their mouths, swallowed, breathed, twitched, wriggled, rolled, stretched, and moved their limbs. One even tried to stand up and run!

Although these movements were a positive sign, demonstrating that the foetuses in those primitive systems were active and seemingly stable, they caused malfunctions in the system; tubes got pulled out with all that wriggling. In fact, for the foetus that tried to stand up and run, the movements cost it its life. The animal sustained massive blood loss through the umbilical blood vessels, from which it had inadvertently pulled out its tubes. Other foetuses perished in the same way.

Swallowing also proved a problem. The foetuses did not drink out of thirst, but rather to help train the muscles of the throat and the digestive system during development – movements that are crucial for survival after birth. (Indeed, swallowing fluid inside the womb is the definition of redundant: a foetus's body-water balance is maintained by the placenta.) In the environment of the artificial womb – a clear plastic tank about the size of a home aquarium filled with yellow liquid – the goat foetuses drank up their surroundings without any care for how much they ingested – and they ingested a lot. Several days after their incubation began, they had accumulated enough excessive fluids that their lungs became swollen. The fluid load also affected their developing cardiovascular system. The scientists were only able to continue the experiment by delivering sedatives and muscle relaxants to the goats, something that most human parents would be reluctant to see prescribed for their own child in an artificial womb, given the potential for life-long addiction.

In the 1980s, researchers in Tokyo were garnering the first promising results in experiments with goats and artificial placentas. Then in 2002, Kuwabara's group developed an incubator consisting of a plastic tank filled with artificial amniotic fluid and a complete artificial placental system to provide oxygen into the blood directly, instead of via tubes into the veins. Unlike previous attempts, which only kept goat foetuses alive for about two days, the scientists reported that the foetuses placed in their fluid-filled plastic boxes stayed alive for three weeks. Like the wobbegongs, they too survived a trial birth.

The artificial womb was becoming a reality.

◎⊙◎

An aquatic environment, an artificial womb, a synthetic placenta: these can surely keep premature babies alive. But could they be used to craft the future of all pregnancies?

Perhaps because of the ethical tangles involved, many scientists working in the field have not disseminated much of their research, or the possible applications of it. That includes scientists such as Hung-Ching Liu, an internationally respected researcher in reproductive biology who, at a conference in 2001, said that her 'final goal is having a child in the laboratory'. And not through old-fashioned childbirth.

By that time, though, Liu had already managed to grow the lining for a human womb, using a sort of scaffolding over which cells, cultured from a woman's womb, could multiply. This 'womb' was only a few sheets of cells in a Petri dish, not an entire organ. But when it was tested using fertilized eggs left over from IVF cycles, the eggs implanted in it at six days, just as they would in a real womb. Liu believes that this approach would ensure that the whole package – embryo and womb – would not be rejected by the immune system when inserted into the woman's body to continue the long process of development.

In the lab, researchers currently are not allowed to grow human foetuses for more than fourteen days, because it is at this point that foetuses develop a neural tube – the precursor of the brain and nervous system. This meant that Liu's experiment could not progress beyond eight days after implantation. Still, going ahead even for this scant time gave her an opportunity to study how the placenta grows, and to see whether she could develop a womb-like device that could remain viable outside of the mother. The device would need to be hooked up to a computer, which would regulate the delivery of liquid to nourish the foetus, the removal of waste products, and the control of the team of hormones that are so finely balanced in the real-life body of an expectant mother. If scientists could achieve this, a baby could conceivably be brought to full term in an artificial womb.

Liu's vision is not fanciful, in terms of motivation or practicality, when you consider two issues. Women – even young women – without wombs are no small minority. In fact, 'absolute uterine infertility', which is defined as a woman's infertility resulting from defective or absent wombs, affects millions throughout the world. In the United States alone, around five thousand hysterectomies are performed in women under the age of twenty-four; nearly nine million women of reproductive age have had a hysterectomy due to conditions including cervical cancer, endometriosis (where uterine cells grow elsewhere in the body, often on the ovaries), and Mayer-Rokitansky-Küster-Hauser syndrome (in which the uterus can be underdeveloped, shaped more like a cord than a sac, or even absent). Most women with uterine infertility have no chance of becoming a genetic mother, except by the use of another woman as a surrogate, and no prospect of ever carrying a pregnancy to term. Liu works with infertile women, many of whom have survived cancer but lost their wombs and reproductive potential to the disease. Her hope is to offer these patients the option of having their own children.

Second, the idea of creating a blood circuit that can serve as a placenta and work alongside an artificial womb and amniotic fluid is complex, seemingly too complex and too dangerous to use in anything outside of the great works of science fiction. But for premature babies, especially those who have difficulty breathing, the ideal situation would be to maintain them in a warm liquid bath, like the womb, attached to an artificial placenta rather than a lung-damaging ventilator. If the conditions in that bath could be set to match the environment of a natural womb, the baby might develop normally, without damage to the lungs and the oxygen-deprived brain. Recently, too, there has been progress in making liquid breathing a reality, through the development of a fluorocarbon liquid with the capacity to carry a large amount of dissolved oxygen and carbon dioxide. The liquid could be inserted into the lung, so that the lung sacs can expand at a much lower pressure, creating an intermediate developmental stage between the womb and life in the open air.

Liu and many other researchers in the field are confident that, despite the complications and difficulties, the technological perfection of an artificial womb is achievable. The French biologist Henri Atlan predicts that, within a hundred years, science will master the complete development of the human foetus from conception. In the meantime, Carlo Bulletti, a professor of reproductive biotechnology at the University of Bologna, says that *partial ectogenesis* – growing foetuses between fourteen and thirty-five weeks of pregnancy – is already within our reach if we were to use all of the knowledge and technology at our disposal.

What would it mean if a foetus could be gestated entirely outside of a woman's body? Ectogenesis is clearly not an ethically easy path for starting or expanding a family. Hand in hand with the creation of a viable artificial womb, doctors and counsellors would have to create something to analyse a number of genetic defects carried by a fertilized egg or early-stage embryo that may not yet be recognized through pre-implantation genetic diagnosis. During in vitro fertilization, some embryos may fail to implant in the natural womb because of random or inherited genetic mutations or those that accumulate with age. With an artificial womb, that process would not work in the same way; the embryos would likely be attached by doctors to the synthetic placenta (or other filtration system that might provide nutrition) meaning that implantation would succeed where it would fail in a natural womb. And the embryo with potential abnormalities might then be able to develop to term in such a highly regulated environment. Would 'pulling the plug' on that foetus in an artificial womb be seen as an early-term abortion or euthanasia?

There is a flip side to that dilemma, of course: it just might be easier to make genetic corrections and modifications to a foetus in a plastic box, which is what an artificial womb is likely to be, less its sophisticated controls. Not only would it be easier to reach the foetus, it would avoid the need to operate on the mother in order to get into the womb. For both mother and child, this would make pregnancy a much safer prospect.

◎◎◎

Another point to consider is whether the only role of a mother's womb is to house the developing embryo and provide what it needs to grow. We now know that a woman shapes the genetics of her child through what is known as epigenetics, which refers to changes written over DNA that are environmental and potentially reversible. Epigenetics is the force involved in genetic imprinting, when a chemical group sits on a stretch of DNA and influences whether and when the genes there work – or don't. These influences can be good, neutral, or bad for a child; epigenetics is agnostic when it comes to development.

The inheritance of characteristics through epigenetics is something that scientists have only quite recently started getting to grips with. But it seems that even from very early in life – including when we are in our mother's wombs – we can be influenced by things that we previously thought had no impact, things like how much a woman eats and how stressed she is. There are, for example, very clear epigenetic signatures that mark those whose mothers have lived through famine and poor socio-economic circumstances. One study of a small town in Sweden found that having plenty of food had an interesting effect: the grandsons of men who ate well had a greater risk of diabetes than the granddaughters of women who did – meaning that the sex chromosomes could be involved. A 2008 study of women who were diagnosed with depression in the third trimester found that their infant children reacted to stressful situations by releasing more of the hormone cortisol, which increases blood sugar and helps with metabolism – getting a person ready for the quintessential fight-or-flight response. And it did not matter if the women were receiving treatment for depression; the stress-response trait passed to the infant regardless. Epigenetic signatures have also been associated with being abused as a child, changing both that person's DNA – and possibly also her offspring's.

There are sure to be other traits passed to a baby through the simple fact of being in one woman's womb rather than another's. And since epigenetics is about shaping genetics, sharing genes with the child in your womb may not make a difference when it comes to these effects. It is possible, for instance, that through epigenetics surrogate mothers are influencing the way a child's genetics play out, including elements of the child's personality – that is to say, how the child adapts to its outside environment. Of course, by the same token, an artificial womb may throw no epigenetic influence on to the foetus growing within it. Whether that would be good, neutral or bad, it is far too early to know. Our knowledge of epigenetic influences is too new for us to begin to contemplate what would happen if they were removed from the process.

What would undoubtedly be good for the foetus would be gestating removed from exposure to undesirable chemicals such as nicotine, alcohol, and other drugs that can be absorbed via the placenta when a mother imbibes. During pregnancy, up to fifteen percent of women are believed to use alcohol, and about five percent use illegal drugs. The proportion of women taking these substances decreases as they enter the later stages of pregnancy, but the effects on the foetus are often worse in the early stages of growth. And drug misuse, illegal or not, is known to have potentially disastrous consequences for an unborn child. Heroin, or more specifically withdrawal between heroin use, can lead to spasm of the placental blood vessels, which reduces blood flow to the placenta and lowers birth weight. Benzodiazapines, which are used to treat anxiety and insomnia among other things, but which are also often abused, slightly increase the risk that a baby will be born with a cleft palate; they are also associated with low birth weight as well as premature birth, and can trigger withdrawal symptoms in the newborn. Cocaine is a powerful constrictor of blood vessels; heavy use increases the risk of several serious conditions, including the placenta detaching from the womb, stunted brain growth, underdevelopment of organs and limbs, and even foetal death. Tobacco causes a reduction in birth weight greater than that caused by heroin, and is a major factor in increasing the risk of Sudden Infant Death Syndrome (SIDS), or cot death. Cannabis use does not seem to have a direct effect on pregnancy, but because the drug is frequently mixed with tobacco, the results can be the same as smoking during pregnancy.

Finally, there is humanity's most accepted drug: alcohol. When consumed in large amounts, alcohol results in reduced birth weight. In the most extreme cases, a baby will suffer the effects of so-called foetal alcohol syndrome (FAS): low birth weight, with general growth throughout life being stunted, including the circumference of the head – and consequently the size of the brain. Children with FAS will also exhibit dysfunctions in the central nervous system, including learning disabilities and certain, characteristic facial abnormalities, known as the FAS face. Children with FAS are likely to have smaller head size and eye openings, an underdeveloped jaw, flattened mid-faces and nasal bridges, smooth philtrums (the slight groove between the nose and upper lip will be absent), thin upper lips, and ear abnormalities.

Substance misuse is often associated with poverty and other social problems, with far-reaching effects on health. And a majority of drug-using women are in their childbearing years. It follows that drug-using women may well be in poor general health before they become pregnant, making their wombs less conducive to a healthy pregnancy, even before ongoing drug use and other issues are factored in. The alternative – a womb outside of your own body – may just be a more salubrious place in which to start life.

◉◉◉

One alternative that is already making headlines is transplantation of a new womb, whether a donated organ or an artificially created womb-like structure, into a woman whose own uterus is damaged or missing. While a womb transplant wouldn't get around the inherent dangers to the mother of pregnancy and childbirth, or the dangers to the foetus from a mother ingesting alcohol or other drugs, it would probably provide a healthier environment – especially given the limits that would be put on a woman's behaviour after having undergone transplant surgery to begin with. But although ovaries have successfully been transplanted in humans, womb transplants have only recently been tried in humans, and early operations with dogs in the 1970s proved unsuccessful.

In April 2000, Dr Wafa Fageeh, leading a medical team in Jeddah, Saudi Arabia (where surrogacy is illegal), became the first surgeon to attempt a womb transplant in a human. The recipient was a twenty-six-year-old who had lost her womb six years earlier, after haemorrhaging during childbirth, and the donor was a forty-six-year-old who had been told she must have a hysterectomy because of ovarian cysts. Fageeh's work was innovative, and the transplant was not rejected by the recipient patient – in fact, she went on to have two natural menstrual cycles. This meant that the graft had been properly done, and had been given a sufficient blood supply. But the transplanted womb had to be removed after ninety-nine days, when a clot developed in a blood vessel that was surgically attached to it. Ultimately, the operation could only be regarded as unsuccessful, since it did not result in a pregnancy.

Within three years, however, scientists began to mark their first triumphs transplanting wombs in mammals. First, mice with donated wombs carried to term and gave birth to normal babies. In 2006, Giuseppe Del Priore, at New York Downtown Hospital, performed a womb transplant on a rhesus monkey; though he was able to establish blood flow between the donor organ and the monkey, the animal was given an incorrect dose of anticoagulants and the experiment had to be terminated within a day. Then, in 2009, a team led by Richard Smith, a consultant gynaecologist at London's Hammersmith Hospital, managed to transplant not just the womb, but also major blood vessels including the aorta, in rabbits. Once the transplant surgeries were completed, the rabbits were placed on immunosuppressant drugs, which helped to prevent the donated womb from being rejected. Alas, despite being mated, none of the rabbits became pregnant. On this occasion, it seemed that the trouble lay with the Fallopian tube, which became blocked and could not carry the fertilized egg to the womb.

These early successes have led to some speculation about the possibility of implanting an embryo into a man. One possibility, in the near term, would be to insert the embryo in the abdomen, the equivalent of an ectopic pregnancy – when an embryo attaches to tissue outside the womb, yet continues to develop. Ectopic pregnancies are dangerous – they can lead to haemorrhaging and death – but a handful of cases in women have been taken to a healthy, live delivery via laparotomy, a form of Caesarean section. In 2008, for instance, a British woman, Jayne Jones, gave birth to a son at twenty-eight weeks gestation; the pregnancy had not terminated earlier because the embryo had attached to a fatty portion of the mother's large bowel, ensuring a good source of nutrition, and the foetus was removed as soon as it was discovered to be outside the womb. This was the first successful delivery in the UK of its sort – and thirty-six medical staff attended.

The eminent fertility expert Lord Robert Winston has commented that 'male pregnancy would certainly be possible, and would be the same as when a woman has an ectopic pregnancy… although to sustain it, you'd have to give the man lots of female hormones'. In such a case, the foetus would be implanted inside a hormone-packed man's abdomen, with an artificial placenta attached to an internal organ – such as the bowel. But apart from all the hormones the procedure would necessitate, the problems associated with ectopic pregnancy would not make it an attractive prospect to anyone. To prevent haemorrhaging at birth, for instance, the placenta would probably have to stay intact, attached to his insides, after delivery. This would be risky for his health – the tissue would either grow, almost like a tumour, or detach or rupture and become lethal when it haemorrhages. If men were to carry embryos to term in this manner, they would, by definition, be experiencing an ectopic pregnancy – which is known to be dangerous to women, and tends to be terminated as soon as it is discovered.

Womb transplantation would be a different prospect entirely – particularly in women. The womb, of course, is a defined space provided for the foetus; as we've seen, it is where the placenta embeds itself, offering a line of communication between the mother and the foetus, not just resource management. And while there are several major hurdles to overcome before the procedure could be considered ready for regular trials in humans, optimism reigns. In 2011, for example, Eva Ottosson, a fifty-six-year old mother of two from Nottingham, England began proceedings to have her womb transplanted into her twenty-five-year-old daughter, Sara. Sara was born lacking a uterus and some parts of the vagina, yet wanted to experience pregnancy and childbirth. In an interview with the *Telegraph* newspaper, Sara expressed no uneasiness about receiving the womb that had carried her to term. 'I'm a biology teacher, and it's just an organ like any other organ,' she said. Eva had asked, 'Isn't it weird?' – but her daughter had answered with an unequivocal no. On the other hand, many people undergoing organ transplants later report feeling as though something about them has changed – not just that a physical bit has been grafted into them, or that they have recovered their health, but that they have acquired new tastes, behaviours, or personality traits, which they usually link to the donor. It might be that the womb, because it has been viewed historically as a vessel for another life, doesn't trigger the same feelings in transplant recipients. But if it does, there may be some odd feelings after the procedure, despite Sara's sure answer.

In any case, the surgery, scheduled for 2012, is not something that the mother and daughter take lightly. Sara noted that she was 'more worried that my mum is going to have a big operation.' Indeed, Mats Brännström, the surgeon planning the groundbreaking transplant, has been working on the procedure for years. He is convinced that it will be more technically demanding than a kidney, liver, or heart transplant. He is especially focused on the complicated connections between the womb and the blood supply and between the womb and the vagina. Will these surgically created connections be strong enough to survive the strain of pregnancy?

Brännström has had successes with some early operations, conducted in sheep. He and his team were able to remove the wombs of five ewes, keep the tissues alive outside of the body for a couple of hours, and then replace the wombs in the original animals, reconnecting the blood supply and the vagina successfully. And four of the five ewes subsequently became pregnant. Brännström and his colleagues have also performed the procedure on mice, rats, and baboons, with two out of five baboons that underwent the surgery resuming regular menstruation afterwards. These are small, incremental steps, but transplantation in humans is the end goal.

◉⊙◉

Of course, even if Brännström succeeds, womb transplants may not be a viable option for everyone – think of a woman who has already had an invasive hysterectomy in order to remove cancer then choosing to undertake a series of transplant operations, with all of the medical risks that would entail. A safer, more desirable course of action might be to turn to a womb outside of your own body. Though today the technology is quite limited, researchers in the field are right to believe that a fully functional artificial womb will come to exist in the next decade or so.

There is obviously a complicated relationship between

an embryo in the womb and its mother, in terms of how a developing baby develops an immune system and takes on board a range of environmental cues while in a mother's body. Indeed, there are many issues that are still not understood, about epigenetics and more. Yet, much has been learned about the underpinnings of disease in the last couple of decades, and that knowledge is breaking open the last remaining barriers to an artificial womb for humans. An artificial womb, after all, will primarily be used to bridge the gap between the fertilization of an egg in a test tube and the movement of the developing embryo into an incubator – since Amillia Taylor's birth a period approaching a brief twenty weeks. And it could help to save pregnancies, whether their origins are in vitro or in vivo, in which the embryo is not yet able to survive with current incubator technology – including many ectopic pregnancies that could endanger the life of the carrying parent.

But an artificial womb could also offer solutions, much as IVF did, both for those with clinical need (which would include gay men if you consider that neither partner will have a womb of their own, and will clinically need one if they want to have a child) and for those who opt for it for various other reasons. For many women who use IVF to become pregnant, the time, pain and expense are wasted when their babies fail to implant in their own wombs. The reasons why this happens are currently not clear, but having access to another womb in a controlled environment certainly sounds like a helpful option for them. During labour, the birth canal is sometimes a treacherous place for babies and the ordeal can lead to death – a scenario that would be avoided if gestation were not inside the woman's body. And because, of course, a woman would not technically have to carry her child, and as pregnancy poses a risk to the mother – in particular, it can genuinely endanger the health of an older mother – this is one advantage, and a use of the technology that becomes very tricky to argue against.

But if you remove a foetus's development from the context of the 'natural' womb, an idea that some opponents say is like putting a foetus in a box for forty weeks, will you also remove the 'special bond' that forms between a mother and her child? To all intents and purposes, however, this question is a red herring: carrying a baby has never been a prerequisite for loving one's baby or being able to bond with it – otherwise the same issues would be an argument against adoptive parents, mothers who use surrogates, and even fathers. In fact, being able to watch, in plain sight, the fragile, doll-like foetus as it develops and grows may encourage a new and special bond. Over the past thirty years, sonograms and other scans have become a regular part of prenatal care, and this ability to view the foetus, as an independent being, is thought to contribute to a maternal–foetal relationship forming much earlier in development – weeks earlier than a mother is usually able to feel kicks and other movements. If it is also the case that bonding is proportionate to the degree to which a child is wanted, parents who have put themselves through any of the gruelling aspects of assisted reproduction – including the artificial womb of the future – may see their great desire to bring their child into the world translate into a great bond with their child, no matter the womb it developed in.

◎◎◎

Not all babies are wanted, of course, which means that an artificial womb or a transplant into a 'willing womb' raises other thorny issues. Such as, if a baby could be made viable from day one using some newfangled contraption, where would that leave the abortion debate?

When a woman has an abortion, she is exercising her right

to remove an unwanted pregnancy from her body, and quite a number of women exercise this right every year. In 2009, there were 189,100 abortions in England and Wales; in 2005, 820,151 were reported in the US. Around forty percent of terminated pregnancies are aborted for medical reasons related to the developing foetus, including the risk of potentially serious disabilities, for example, of damage to the nervous system or Down syndrome. (An estimated ninety-two percent of all women who receive a prenatal diagnosis of Down syndrome choose to terminate the pregnancy.) The remaining sixty percent of abortions are chosen for reasons related to the mother – her own physical or emotional health or her relationship to the father, among many other factors. It should be stated that a woman's right to an abortion does not give her the right to kill a child; rather, the aim is to end a pregnancy. This is why, under the law, we consider the foetus to be a collection of cells, not a baby, until some demarcated point when the cells could live on their own outside the womb.

Before birth, the rights of babies – that is, foetuses – are not protected; under current UK, Canadian, and US law, foetuses have no rights at all. In a handful of cases, however, American mothers have been charged with child abuse for behaving in ways that allegedly harmed the foetus they were carrying. If an artificial womb were created in which a healthy foetus scheduled for abortion could survive to term, the issue of whether it should be nurtured there would become a matter for politicians and public policy to decide. Ninety-one percent of the abortions performed in the UK in 2009 were conducted when the foetuses were at thirteen or fewer weeks gestation – too early for today's incubators. If they could conceivably be kept alive, would medical staff have an obligation to resuscitate them and place them in an artificial womb? Would it be better for society if these pregnancies were not aborted, if the embryos survived to become people with inalienable human rights?

Already, the relatively antiquated incubators in modern hospitals have proved to be an ethical minefield when conflicts arise between the desires of premature babies' parents and the obligations of medical staff. One such battle began on 21 October 2003, when a baby who had only been in the womb for twenty-six weeks was born in Portsmouth, England. Tiny Charlotte Wyatt was only 12.7 centimetres (five inches) long at birth and weighed 458 grams (sixteen ounces), instead of the average 3.5 to four kilos (approximately 7.5 pounds) for a full-term baby. Charlotte was fragile; her organs – especially her lungs, heart, and brain – were extremely underdeveloped. She nearly died after delivery. After being resuscitated, Charlotte suffered severe brain damage and several of her organs failed; she was left blind and deaf, her kidneys were compromised, and her lungs were so severely injured that she required a constant supply of oxygen. The extreme immaturity of her body also meant that her immune system was unprepared for the world outside the womb, and any small infection could be lethal. There was little hope of her living beyond childhood.

Charlotte's team of doctors contended that, however long she lived, she would not only need continual medical attention, but would also likely be in constant pain and experience a life of extremely poor quality. Medical opinion was weighted in favour of no longer resuscitating the child when she next suffered cardiopulmonary failure – as she had on three occasions – and instead allowing her to die with some measure of comfort. The medical staff argued that with every resuscitation they performed, Charlotte's lungs became increasingly delicate, and aggressive treatment was not in the child's best interests. Prolonging her life, in fact, appeared to constitute an assault of 'inhumane and degrading treatment' under Article 3 of the European Convention on Human Rights, as the potential long-lasting harms to the person would ultimately exceed the benefits. The doctors argued in favour of palliative care alone.

Charlotte's parents disagreed. As committed Christians, the Wyatts believed that their daughter's life should be preserved at all costs. So when doctors refused to resuscitate Charlotte for a fourth time, her case was brought to court. The doctors won the legal right to let her die, should her body shut down again. Yet, Charlotte did not die, as expected, and the 'do not resuscitate' order was eventually lifted, in 2005, when her parents showed that Charlotte was no longer in constant pain or unable to respond to stimuli.

Over those two years, however, the extreme stress of the situation had led to the breakdown of the Wyatts' marriage, with the two parents visiting their severely disabled child only infrequently in the hospital ward. A second series of legal clashes ensued, this time over sustaining care for Charlotte. In the end, the child was placed with foster parents. By 2009, her father was visiting her monthly, according to an interview with the *Daily Mail*. He reported that, though Charlotte still needed some oxygen every day, she loved to listen to nursery rhymes and could stand and walk with the help of a walking frame. 'Going through the courts to keep Charlotte alive totally drained me,' he said. 'But now, when I look at her smiling face, I know it was the best thing I ever did.'

While it is currently feasible to keep the very premature alive, good health and quality of life are by no means guaranteed – and there can be a devastating toll on both child and parents. An artificial womb that can sustain and continue the development of extremely young foetuses could completely reinvent the parameters of neonatal medicine, helping to give children like Charlotte a less traumatic life.

⊚⊙⊚

Regardless of such gains, a fully functional artificial womb will also present entirely new ethical dilemmas, including some we may not be ready to negotiate. What if a foetus that would otherwise be aborted could be removed from its mother's body and gestated artificially? Would that improve the chances of adoption for a child, given that many couples prefer to adopt a baby rather than an older child? Would each year's 189,574 aborted pregnancies, as occurred in 2010 in England and Wales, be viewed as the prospect of a joyful miracle in the tradition of the first test-tube babies, or would they be seen as supplanting the placement of older children needing a home?

How will this new technology alter the identity of a mother, a role that would cease to trigger a biological bond, even if her own egg is used? For instance, there has been a great deal of research into the hormones oxytocin and arginine vasopressin. In mammals, the levels of these hormones are elevated in mothers' brains. Oxytocin levels also increase during labour and reach a peak at the time of delivery. Both oxytocin and vasopressin have been linked to the instinct towards maternal care and mother–child and other affectionate, family bonding. The hormones have even been seen to rise when mothers engage in other supportive and bonding behaviours, long after pregnancy, though it is not known how and why this occurs. If a mother did not experience the increase in hormones related to pregnancy, would it make a difference later in life? Would it be possible to give a mother a dose of the hormones, in place of this natural release? It is apparent, from the experiences of many adoptive mothers, that a mother–child bond forms even in the absence of pregnancy, but it may be that those who choose to adopt happen also to have a strong instinct for maternal care.

It may be that separating the physical experience of pregnancy from the body of a mother also requires separating it from the mother's biological brain.

Further, since a child's identity is in part shaped by the communication of hormones and other information from mother to foetus, pregnancy via an artificial womb would redefine what it means to be a biological parent. Perhaps in the future a mother who uses an artificial womb will primarily be seen as a *genetic* and *social* parent, since all of the biological exchanges of pregnancy will gain new significance. Could the artificial womb become yet another symbol of the ways in which a woman is or is not a 'good mother'? By relinquishing the chance to shape her child's development from embryo to full term, a mother might be ensuring a more resilient temperament for her offspring, after all. In a case where a woman uses a donor egg and an artificial womb (by choice or necessity), the baby will have neither gestated with the mother nor bear any of her genes. Would the egg donor have more legal rights to the child in this case? In these ways, the very concept of an artificial womb reveals how societies view women. Even in the twenty-first century, a woman is still often defined by her role in procreation.

Consider, for instance, surrogacy, the practice of using another person's womb to carry your embryo to term. The role of surrogate mother, sometimes described as putting up a 'womb for rent', is considered by some to be exploitation, especially as the practice has been more and more often outsourced to countries where a high proportion of the population live in poverty. Countries such as India.

Since 2002, when the Indian government legalized paid surrogate pregnancy – critics say they did so in the hopes of giving birth to a new 'pink-collar' industry – young Indian women have been queuing up to become surrogate mothers. There are doctors in nearly every major Indian city working with women who want to be surrogates; there is even a town in the state of Gujarat – its name is Anand, which in Sanskrit means 'bliss' – that is poised to claim the mantle of the nation's go-to centre for paid pregnancy. In 2009, one Mumbai doctor told the London *Evening Standard* newspaper that she delivers more than fifteen babies for British couples every month – about one every forty-eight hours. (Unfortunately, despite the legalization of the service, the government does not keep reliable numbers of how many women have become surrogates.)

It's not surprising that Indian women are signing up in hordes – they are paid between $6000 and $10,000 (£3700–£6000) to be a surrogate, which amounts to about fifteen years' wages, on average. The rise in infertility in industrial nations is certainly fuelling this 'business', as commercial surrogacy is banned in most of Europe and in many US states. Couples, most commonly from the UK, US, Germany, Taiwan, Japan, and Australia, go to India to take advantage of these services, because, even with the travel costs, it will cost them just one third of what it would in their home countries.

There are complications to this outsourced labour. Women in India are sixty-nine times as likely to die from childbirth-related issues due to inadequate access to good medical facilities. The Indian government has not put in place any regulations to protect the rights of surrogate mothers. As it stands, surrogate mothers are looked after during their pregnancies, but they receive no compensation for medical difficulties that arise after childbirth. These women are at risk of long-term liver problems – a side effect of being pumped full of the hormones used to prepare the body for pregnancy. They also may face the common complications of pregnancy: the risks of toxaemia, anaesthesia, and haemorrhage, to name but a few. Further, it has been documented that many couples who have returned from using surrogate services in India have delivered twins. Multiple births generally mean lower birth weights for the babies and more dangers that arise to the mother during childbirth – so much so that implantation of more than one embryo during IVF is frowned upon by the National Health Service.

Plus, we just do not know what are the true risks of carrying a child to term who has no genetic relation to you. We do know that a mother who has been exposed to a partner's sperm before she conceives his child is less likely to suffer from pre-eclampsia, a potentially life-threatening condition in which blood pressure and urine protein levels soar. Pre-eclampsia may be related to immune recognition, that is, when the mother's immune system antibodies, after being exposed to the father's foreign antigens, allow the placenta to penetrate the wall of her uterus more deeply. Researchers have found that the many genes that control the growth of the placenta are expressed from only the father's DNA. This could mean that the growth of an embryo and its supporting placenta in the body of a woman who has never been exposed to the genetic father's antigens, *and* who herself has given no genetic input into that embryo, may be up against an as-yet-uncatalogued threat to her immune system – as well as that of the foetus she is carrying.

There are also looming issues unrelated to health. In one recent case, a Japanese couple who had paid an Indian surrogate ended up divorcing, and the ex-wife no longer wanted the baby – who had not yet been born. The surrogate mother didn't want the baby either, and under Indian law, she was prevented from handing over the child to the father. After much legal wrangling, the paternal grandmother was given custody of the infant.

Surrogacy in India is a lucrative business, and family hierarchies in the country still hold great power – especially over their female members – which raises the question of whether all of the women caught up in the system are truly doing so out of choice. Could some families be putting pressure on their young women to join the ranks of surrogate mothers in order to benefit household economics? One family, for instance, was recorded to have three sisters pregnant as surrogates at the same time; their sister-in-law was pregnant with her second surrogate child too. Likewise, many surrogate mothers live in houses that have been described as akin to a fertility reality show. For the duration of their pregnancy, up to fifteen expectant mothers may be packed into a house, where they are overseen, Big Brother-style, by a former surrogate mother.

A doctor who implants embryos in surrogate mothers at a prominent Mumbai clinic reported to the London *Evening Standard* that business is very fertile indeed. 'Surrogacy is spreading at a very fast pace here and there have been very few complaints,' he said. 'Our email inquiry box is full of messages from people from all over the West.' Another fertility specialist at the clinic emphasizes the convenience in his pitch: 'There is no paperwork involved; the couples don't have to go through any lawyers; it's a clean issue – and there is no litigation.' While such loopholes may be attractive to the doctors' relatively wealthy clients, the Women's Protection League of India disagrees that surrogacy is a positive development for the surrogates themselves, especially with respect to their health. A spokesperson for the group said, in no uncertain terms, 'This is exploitation and I totally condemn surrogacy.'

An artificial womb could be the great equalizer for women – a way to end the exploitation of another woman's body in order to bear a child when one woman discovers that her own body cannot do so for her, or even if she decides that it's simply not convenient to do so. It would mean that a woman's big life choice would be *whether* she will bear her child, rather than *when* she might do it. And this liberated mother could carry on with her life as usual up until the moment of birth, much as most fathers do.

The invention of a human artificial womb might also mean that the divide between mother and father can be dispensed with; a womb outside a woman's body would serve women without wombs, transsexual men, and male same-sex couples equally without prejudice. For this reason, some feminists have argued that the quest for the artificial womb comes from a deep-seated desire to displace women and dissociate birth from the maternal body – effectively, to erase the mother. And in a case of fact being stranger than philosophical fiction, an internet forum for fathers campaigning for parental rights when marriages dissolve has seen messages advocating for an artificial womb – because it would free fathers from the tyranny of those mothers who keep men apart from their children.

The cultural divide between mothers and fathers appears to be closing, at least in some parts of the world. Two generations ago, fathers were not as hands-on and engaged with child-rearing as they are today. There hasn't been a change in the biology of sex in that time; the change has come through our culture, including the tools available to us to equalize the distribution of labour (in the sense of work). When an artificial womb becomes available, an equal distribution of labour (in the sense of childbirth) will finally be within reach. This will mean that women will be freed from the dangers of pregnancy and will be able to work productively throughout gestation; it will also give men an essential tool towards being able to have a child entirely without a woman, should they choose. But it also means we will have to consider the most basic questions of gender: why are the roles of mother and father still seen as different to most people on the planet? Why can't a man be a 'mother'? Why do we care so much about what it means to be a 'mother' rather than to be a 'parent'?

By all reasonable estimates, in the near future we will conquer the tyranny of time and the tyranny of the womb. The question remains if we can also conquer the tyranny of human prejudice too.

9

GOING SOLO

There are plenty of reasons not to put up with the world as it is.

José Saramago, interview with the *Guardian*, April 2006

The UK Office for National Statistics' 2012 study of lone parents with dependant children reports that the traditional family household of a married couple with a child or children is now three times less common than it was just a generation ago. Families headed by only one parent comprise twenty-five percent of households in the UK and twenty-eight percent in the US, and in the US, the so-called nuclear family now accounts for fewer than twenty-five percent of households, compared to forty percent in 1970.

The majority of single-parent families are created by circumstance – separation, divorce or the death of a partner. Recent decades have also seen the rise, however, of the solo parent, a name used to distinguish these single-parents-by-choice from other single-parent families where a two-adult household has been broken apart, often with economic consequences, and for that reason is often associated with disadvantage and, sometimes, pity.

Not so solo parents, who are generally, at this moment in reproductive history, single women in good financial circumstances who are approaching or just past the menopause, and who have made a conscious decision to use advanced technology to go it alone. These women have not necessarily experienced infertility problems; instead, they turn to assisted conception techniques in order to have a child of their own without a partner. The families that result are not always a solo mother and a solo child alone. Some solo parents want to give a sibling to an only child who may or may not have been the product of a partner's sperm.

Because of their age, many solo parents must use more than one reproductive technology when they decide to have a child. A woman nearing menopause and lacking a male partner will, for instance, usually need a donated egg, donated sperm, and in vitro fertilization to bring the two together and then successfully implant them in her womb. On some levels, such a pregnancy is natural – it's just that all of the bits and pieces are happening outside the usual conception. The family that results isn't related genetically, but its members are related *biologically*. The births mimic everything that happens when a man and a woman have a child. In some cases, the solo mother might try to find donors that will provide genetic material that, on the surface, appears to be their own, making it impossible for the person on the street (or at the nursery) to distinguish a solo-parent family from more 'traditional' types. Indeed, while most solo mothers say they plan to tell their children that their father is a sperm donor, many admit that they probably will not tell their children that the egg that made them was donated too. We still put a great deal of emphasis on the meaning of that genetic contribution, after all, and the desire to become a parent is wrapped up in those definitions.

How does this affect the welfare of the children? So far, research has focused on the effects of growing up fatherless, and most of it involves households where a single-parent family has been created by force, not by choice. Yes, there are negative consequences for a child's cognitive, social, and emotional development when economic hardship, parental conflict, or parental death is part of the family story. But these effects should not be generalized to include children born into solo-mother families. From the early accounts, children of solo parents experience neither trauma (from the break-up of the family) nor financial hardship. For many, in fact, the situation is quite the opposite.

Although there have been few studies of solo parents, a remarkably consistent theme emerges in the interviews that have been conducted: the women (and solo parents are almost always women, given the limits of today's technology) realized that they had no other viable options for becoming a mother. Time was running out, and they had no long-term partner in their sights. The risk of having a child through a casual sexual liaison was too great, be it the chances of contracting a sexually transmitted disease or of being deceived about the other person's intentions towards any child that might result. A few said that they simply wanted to have a child without the involvement of a man.

Of course, this all assumes that a woman is seeking to become a mother biologically. As it happens, most of the solo mothers studied mentioned that adoption was their first choice – they described it as a 'more moral' solution to the desire to start a family. But adoption wasn't a viable option either. In the US and the UK, most adoption agencies prefer to place very young children with couples in their twenties or early thirties; older, single women reported that their only real option was to turn to international adoption bureaus, which charge at least

£20,000, or $32,000, for their services. In contrast, buying an egg or sperm costs ten times less. Given that raising a child involves a considerable amount of expense, spread over twenty or more years, this financial decision could be interpreted as a first signal of parental responsibility.

In at least one recorded case, the decision to become a solo parent was tightly bound with the woman's religious beliefs. The woman, part of a small group study of solo mothers, said that having a baby using medical interventions was her only moral choice since she wasn't married, and her religion considered sex outside marriage to be a sin. She effectively took the scientific route to a church-sanctioned virgin birth.

⊚⊙⊚

While buying eggs from a younger woman is a highly effective way for an older woman to have a child, having to use donated eggs is not ideal – especially for the donor. Donating eggs is an invasive, inconvenient, and, at times, painful process, requiring daily hormone injections and the extraction of eggs from the ovary. Truly altruistic donations of eggs are few and far between. As in the case of wombs-for-rent, there is scope for abuse. Women from poor socio-economic backgrounds may submit to – or be coerced into – successive egg donations to make money, or even to gain access to fertility treatments for themselves. This phenomenon has been called 'fertility tourism' or, in its more nefarious forms, 'egg trafficking'.

Since the first successful birth from a donated egg in 1984 at the UCLA Medical Center, the trade in eggs has grown exponentially – and offers a classic study in free-market supply and demand. Originally, egg donation was developed as a therapy for young women with premature ovarian failure; only more recently has it become more widely used as a means of overcoming the age-related decline in fertility. In the early days, egg donation as a charitable act was the norm, but as the demand for eggs has increased – because of both technological and social change – the motivations for supplying eggs have shifted. In the UK, the egg-donor market shows just how few egg donations are motivated purely by altruism, since very little money is paid to a donor for her eggs, and very few eggs are donated. Donors are not paid for their services beyond their 'expenses'. For some, this is enough money to make the effort worthwhile, but the donation rates in the UK are lower than in places where women are paid more. Further, egg and sperm donors are no longer permitted to be anonymous in the UK, which discourages some who do not want to be faced with a genetic child eighteen years down the road. In France, the most restrictive of European countries when it comes to egg donation, the eggs must be deemed to be a completely free gift from one woman to another. In one recent year, just 144 French women volunteered to undergo the donation procedure. There is no financial compensation in Britain either, but in the same year, 1509 women donated eggs in the UK (although 999 of them were sharing eggs while undergoing fertility treatments themselves).

In other countries, where it is legal to pay vast sums of money to egg donors, the fertility clinics are in demand. Spain, for example, boasts more private IVF clinics than any other country in the world – and the country's clinics also claim to get the best results. Good enough that French women flock there to receive donor eggs, since so few are available at home. If you undergo IVF in Spain and use 'donated' eggs instead of your own, you will probably pay an extra £2000 for your treatment, which gives some sense of what these eggs cost to 'buy'. There's a very good reason for the wonderful success rates: the Spanish clinics advertise to young women, at the peak of their fertility, and offer lucrative compensation – around four times the UK rate, from £800 to over £1000. Spanish law allows payment for the donor's time (as compared to her eggs), which skirts EU regulations to avoid exploitation of reproductive material. This is the case in Cyprus, one of the other hotspots for continental egg donation procedures. And in some US states, it is quite legal to pay young women huge sums of money for their eggs. Young women are often recruited through private clinics or online agencies; a Google search will bring up tens of thousands of results. But especially desirable donors – usually university-educated women, offering the 'right' geneticmake-up – are also targeted through ads posted at university campuses and in student newspapers. One fairly typical notice reads:

> Egg donor wanted – $35,000 compensation. We are a couple seeking a high-IQ egg donor to help build our family. You should have or be working on a university degree from a world-class university; you should have [high] standardized test scores and preferably some outstanding achievements and awards.

When an ad isn't lure enough, there's always the alternative of seeking help from the universities themselves. The medical school at the very prestigious Yale University, for one, runs an egg donation programme under the legalistic label 'third party reproduction'. Yale is, after all, a member of the Ivy League, and according to a 2009 report in *Marie Claire*, payment for the eggs of blonde, blue-eyed, athletic undergraduates have sold for as much as $100,000. That's an attractive sum for a student who is likely to be staring down a massive debt bill for her higher education. At Yale, tuition fees alone now stand at $40,500 per year – which adds up to roughly $160,000 over the course of a standard four-year undergraduate degree.

In the US, donor eggs generally come from American women, but in Spain and Cyprus eggs may come from women who live anywhere in Europe, so that women from places that are experiencing tough economic times may travel quite a distance to make some much-needed cash. One Eastern European woman, who decided against donating her own eggs but witnessed many others do so, told the *Observer*: 'They work the cabarets, they'll sleep with men, they'll sell their eggs, and then they go back again.' She seemed to equate each of these activities – reflecting on the various ways in which women, desperate for money, may try to earn a living. And since women are highly unlikely to be sleeping with men and selling their eggs simultaneously, you almost have to assume that egg donation might be an escape from the other. In Russia, the £800 often paid by a Spanish clinic for one cycle of egg donations equals a year of average wages. The *Observer* noted that one clinic even offered a $500 'signing' payment to women willing to be flown from the Ukraine to Cyprus for egg-donor screening.

Because this is a 'free' market, clinics offer all sorts of bonuses. If a woman is willing to donate more eggs, she can earn an extra fee – but producing more eggs means taking more hormones, twice the dose that is recommended. This can be very bad for the health. Premature menopause, uterine cancer, and ovarian hyperstimulation syndrome can result. The therapy has also put several women at risk of death. In one case, a Stanford student who had agreed to donate eggs for a fee of $15,000 experienced a rare adverse reaction to one of the fertility drugs she was given. The side effects were devastating. She suffered a massive stroke, which left her in a coma for eight weeks with long-term brain damage.

⊙⊙⊙

The money paid to the sellers of eggs is probably dwarfed by the revenues that stream into the 'middle-man' clinics. Because these private clinics are able to pay top price for eggs, they have little to no donor shortage and are also able to perform what some have called 'personalized baby marketing': if the client pays a fee over and above what the clinic has paid to the donor, then the client may select a donor based on her height, weight, eye colour, educational attainment, and other criteria.

In an effort to curtail the influence of money in egg and sperm donations, in April 2005 the European Parliament adopted a resolution banning trade in human cells and embryos. The legislators were moved to act after reports emerged that a clinic in Romania was sending 'mail-order eggs' to the UK, with the UK government proposing to pay up to £1000 as an incentive to entice more donors in the future. While compensation for donor expenses is allowed under the European resolution, the regulatory body responsible for administering the rule stated that a payment in the range of £1000 would be well above the allowed limit.

Despite the new law, the market continued to prosper. In Romania, five people were arrested and held in detention in the summer of 2009 over suspicion of trafficking human eggs. The chief prosecutor of Romania's organized crime department also held the suspects on broader charges, including allegedly practising medicine without a permit and being involved in a criminal group. Before their detention, two of the group, gynaecologists from Israel, had run an IVF and plastic surgery clinic in Bucharest. The gynaecologists were suspected of recruiting women aged between eighteen and thirty and paying them around £150 ($300) for their eggs. The eggs were then sold on for £5000 to £7000 ($10,000 to $14,000) to clients from Israel, Italy, and the UK. According to the Romanian newspaper *Gardianul*, the clinic's annual earnings were around €20 million (£14 million, or $25 million). The case closely followed the arrest, on similar charges, of thirty Israelis who worked in a separate fertility clinic in Romania, which suggested that the scheme was not an isolated case of a few bad apples taking advantage of an otherwise finely controlled system.

Then, in 2010, a fertility clinic in Cyprus came under surveillance after authorities received claims of human egg trafficking. The clinic, run by mostly Russian staff, relied primarily on donors from Eastern Europe. Three Ukrainian women in their thirties, all of whom were living and working legally in Cyprus, were questioned after donating eggs to the clinic, and said that they had received more than reimbursement for their expenses. According to media reports, the women claimed to have been paid €1500 (about £1000, or $1900) for their services, though the police would not confirm the figure. Officially, the clinic was shut down in May 2010 on orders of the health ministry for failing to provide full data for the provenance of embryos, eggs, and sperm. This effectively meant that the donors were untraceable. Indeed, Cyprus had become a favoured destination for the egg trade because of its clinics' low prices and donor anonymity. The police investigation into illegal egg trafficking had to wait for the Ukrainian government's approval.

The free movement of people across the European Union makes it difficult to crack down on the trade in eggs, all the more so since the demographic changes over the past few decades mean that some countries have a shortage of eggs and others have a youthful supply. Jacques Testart, the research director at INSERM, a medical institute in Paris, was not particularly surprised by the stories coming out of Romania and Cyprus. 'There are rumours circulating about trafficking in Europe, although they are difficult to prove,' he told the news agency AFP. 'There will always be a need for the "hens" and there will always be women who do that to earn a bit of money… especially in the current economic crisis.' One gynaecologist, who spoke to AFP on condition of anonymity, claimed that egg trafficking is common in Cyprus. 'Everyone knows that, but we don't do anything [about it],' the doctor alleged.

Eric Blyth, a professor in the Department of Human and Health Sciences at the University of Huddersfield, has identified three key characteristics of the countries that have become popular destinations for fertility tourism: 'First, the lack of regulation affording adequate protection for the parties most directly affected, i.e., donors, surrogates, patients, and children; second, the operation of a commercial market in human gametes – especially eggs – and women's gestational services; and third, a level of secrecy that helps to conceal unprofessional, unethical, and illegal practices.' Because of these issues, the UK's Human Fertilisation and Embryology Authority (HFEA) has called the use of foreign egg donors a 'profoundly exploitative and unethical trade'.

There have also been shenanigans involving sperm acquired by ill-gotten means. In the past decade, a number of internet businesses have cropped up that claim to be able to put people desperate to become parents in touch with potential sperm or egg donors, or to supply donations directly, serving as a middle man. The online services may appear to be an easier, cheaper, and less bureaucratic option than going through a government-licensed clinic; they also rarely advertise, say, the sperm shortages that characterize the market, which makes them seem more likely to deliver the goods. But such sites may pose a risk to people trying to find help. In fact, since April 2007, it has been unlawful to 'procure, test or distribute' human eggs or sperm for human reproductive use in the UK without a licence from the British government authority that regulates fertility work. Regulated clinics in both the UK and the US are required to freeze and store sperm for six months before it is used by a woman, during which time the clinics test it for HIV and other diseases, but internet traders selling 'fresh sperm' had not been required to do such checks. The law, called the Human Fertilization and Embryology Act, was brought in to regulate the use of fresh sperm, with a view to ensuring it is safe. In addition, via the internet there is generally no way to confirm that the donor is who he says he is, and, as a result, the safeguards that UK and European law offers to parents and any resulting children may not apply. And given the internet's ability to cross borders, there is little that can be done to force a site based in another country to follow local law – an issue made even thornier in the US, where state laws may vary considerably.

In 2009, the UK saw the first prosecution of an internet sperm trader under the new law. The case involved a website called Fertility First, through which fertility patients could select from a database of anonymous sperm donors and order 'fresh sperm' to be delivered, for a fee, direct to their front door. A customer, Melissa Bhalla-Pentley, paid £530 to receive this convenient sperm supply, a price tag that allegedly included reimbursement of the sperm donor's expenses as well as a site membership fee, a courier charge, and a per-cycle cost for the sperm itself. When she failed to get pregnant, she arranged for another donation – an extra £300 charge. She had requested the donor's medical records, but when they arrived she noticed that his name was visible, 'just lined through with a black marker'. Something seemed amiss. At the very least, the company had breached protocols of donor privacy. When her request for a refund was refused, Bhalla-Pentley went to the police with her complaint.

When the case came to court in 2010, it emerged that the

entrepreneurs behind Fertility First had earned up to £250,000 from about eight hundred customers. Two men were found guilty of procuring and distributing sperm without a licence, as required by UK law. The sperm donors were reported to have received no payment at all for their services, nor had they realized, by their account, that Fertility First was unregulated. In the *Daily Mail*, reporter Laura Topham related how the enterprise had been hatched after the men overheard a childless woman in a pub talking about her desire to get pregnant – 'she wanted sperm delivered like milk in the morning'.

This is what reproduction looks like when capitalism's invisible hand has a free rein. Where there is demand, a supply will be found.

◎◎◎

In order to give birth, a woman first needs eggs, and then needs sperm. That's how reproduction works today. If she doesn't have good quality eggs of her own, and has no access to safe sperm, there are currently few channels through which she can acquire either, outside of the donor market. But doctors are developing ways to bypass the market – including a treatment that is already being used successfully.

Eggs develop in ovaries, of course, but what is contained in the ovaries are not strictly eggs but immature 'follicles', clumps of cells that contain a single oocyte that grows and develops into a mature, ready-to-be-fertilized egg. This is why the key hormone in timing of conception and in IVF treatments is called follicle-stimulating hormone (FSH): the hormone stimulates the follicles to grow and eventually erupt, releasing the egg during ovulation. This means, however, that acquiring a healthy ovary, or even strips of tissue from a healthy ovary, and transplanting it into a woman is another way of getting around the problem of scarce good eggs. If the tissue has follicles that are still receptive to FSH, a woman would once again be able to generate eggs, no matter what her age. Take the case of Susanne Butscher, a woman who became infertile at the age of fifteen when her ovaries failed, causing her to experience a very early menopause. In November 2008, at age thirty-eight, she became the first woman to give birth after receiving a transplanted ovary.

Ovary transplants were developed for women, like Butscher, who suffer early menopause, or for those undergoing chemotherapy or radiotherapy to treat cancer. While strips of ovary can be removed from a woman without ill effects, removing the ovaries themselves can trigger premature menopause, with all of the associated ill health effects. However, Dr Sherman Silber, who performed Butscher's transplant surgery, sees the potential for using the procedure as allowing women who have delayed motherhood for any reason to improve their chances of having a baby later in life. Rather than freezing eggs and undergoing IVF with them, a whole ovary could be frozen; the tissue would be viable for up to a decade. While the extraction and transplant surgeries are invasive, they could circumvent some of the problems associated with other fertility treatments. Children conceived through ICSI or IVF, even those conceived simply through the use of drugs to induce a woman's eggs to be released for harvest, appear more likely to have problems with genetic imprinting, growth, and defects. And, unlike with IVF, a preserved or new ovary gives women the option of conceiving a child via sexual intercourse with a fertile partner. After her transplant, Butscher started having periods again, for the first time in twenty-three years, and she and her forty-year-old husband used no other fertility treatments. Indeed, the oestrogen, progesterone, and testosterone produced in the ovaries affect the female body in many ways, including protecting the bones from osteoporosis, and Butcher's bone health improved as well.

Not everyone found good news in Susanne Butscher's case. Just as happened when Louise Joy Brown entered the world as the first test-tube baby in 1978, the delivery raised moral and social concerns in many quarters. Chief among them: were surgeons using science as a tool to alter the child-bearing age for women? The UK's Royal College of Midwives, for instance, stated that it would be preferable for surgeons to limit ovarian transplants and other reproductive technologies to women who are 'truly' infertile – meaning that they have become infertile in their twenties or earlier – and who are desperate to conceive. This would include Susanne Butscher, of course, but also survivors of childhood cancers who show evidence of normal ovarian function, but who will require a therapy that would otherwise destroy their ovaries. Right now, women battling cancer at a young age must either become a mother before treating the cancer, or treat the cancer – hardly a happy choice to make. But is it fair to say that some women, because of a medical condition, 'deserve' to benefit from these technologies, while others, because of societal and economic conditions, do not?

Those working on the frontline with people who are infertile argue that modern lifestyles are altering the child-bearing age for women – making it difficult for women to have children earlier in life. And then there is the question of how to define a 'truly' infertile couple. Yes, a man or a woman may be biologically ill-equipped to have a child together because of the health of their sperm and eggs, but a lesbian couple could make the case that they fall into this category, too: they don't have the healthy sperm they need to have a child. In a statement on why IVF treatments for infertile couples should be a priority for the National Health Service, the British Fertility Society wrote that those 'involved in infertility services are all aware that we are not just dealing with a physical pathology. Infertility is a disease, but it also has fall-out beyond that... causing mental health problems, depression, stress-related illnesses, and so on.' These are serious health conditions, and if we have the tools to treat the underlying problem – the inability to have a child, at a time in life when a child is desired – shouldn't we do so?

Susanne Butscher, for one, would probably agree with that idea. She and her husband named their baby Maja, for the Roman goddess of rebirth and fertility, and Butscher said Maja gave her 'a sense of completeness [she] would never have had otherwise'.

◎◎◎

Ovary transplants require a supply of compatible ovaries, so the problems that come with the trade in eggs and sperm – replete with misstatements, privacy violations, skirting regulations across borders, and criminal scandal – may well apply here too. And there are documented reports of a black market in body parts for transplant surgeries, so the infrastructure is already in place for ovary trafficking if the surgery is allowed to go forward on a larger scale. So scientists are considering how to take the idea behind ovary transplants and apply it to reproduction, without the demon of a limited supply of organs to meet demand.

Eggs and sperm are collectively called *germ cells* for their potential, somewhat like seeds, for growth to emerge after a period of dormancy. Early in evolution, a process of segregation must have happened so that germ cells were kept apart from all our other cells, possibly as a way to protect the integrity of their essential genetic material, in case nutrients became scarce and reproduction had to be delayed. Germ cells, of course, have a special place in our life cycle because they are essentially immortal – they provide the fundamental link from one generation to another. They must also be able to re-create the entire organism. At the same time, as we have seen, they have a pernicious shelf life, made all the more frustrating to many people by the fact that each of us has a limited supply. The first question to tackle, then, is whether a woman's body might somehow be triggered to make eggs on its own, without resorting to getting follicles from an outside source – in essence, creating a new supply of eggs. In fact, there is a debate around whether a woman is really born with all the eggs that she will ever have in her life, a debate that has been raging for nearly a hundred years, since the German anatomist Heinrich von Waldeyer-Hartz first proposed, in 1870, that all female mammals stop producing eggs around the time they are born.

'Less evolved' species, such as flies, birds, and fish, can and do generate new eggs throughout their adult life, which means that the ability to produce new eggs was lost somewhere during evolution, before the emergence of mammals. Yet the ability of males to produce sperm throughout adulthood was conserved in species from flies through to man. Why would the process of evolutionary selection deem it an advantage to endow women with only a fixed number of eggs that sit stagnant and are subjected to years, if not decades, of ageing before they erupt from a follicle during ovulation? The more logical, more robust approach would be to keep generating fresh eggs and fresh sperm. At least in theory. But theory aside, new evidence points to signs that mammals also have the potential to generate new eggs.

Stem cells have received a great deal of attention in the press for their ability to renew on their own, unlike all of our other cells, which only age and die off. Every type of cell, tissue, and organ in our bodies must be created from the genetic material contained in the egg and sperm, the precursors of the embryo. This power to generate the totality of components required to build a human is called *totipotency*. Stem cells derived from embryos are better suited to fulfil this role than those taken from an adult body, because in early embryos every cell has the potential to generate into any number of different cell types. As development proceeds, and cells become more fixed and decided as to their ultimate fate – that is to say, after a certain point in development, one cell will only be able to become a brain cell, another will only become a muscle cell, and so on – the stem cells lose their totipotent potential and become *pluripotent*, able to generate several different cell types but not all. But because extracting embryonic stem cells currently requires destroying an embryo, the technique is besieged with controversy, particularly in the US. Whether it will become possible to coerce adult stem cells into acting in a similar manner has still not been proven, though researchers are trying to revert skin stem cells to an embryonic state.

Regardless, egg-making stem cells have been found in the ovaries of adult mice, monkeys, and humans that retain stem cells with the capacity to renew the egg pool. At the moment, there's still no evidence that these cells form new eggs naturally inside a woman's body, but experiments are being conducted to see if they could be coaxed in a dish to make eggs. And if they could be coaxed to do the same inside a woman's body, these stem cells could provide an unlimited supply of eggs, and also be used to postpone age-related ovarian failure and perhaps the menopause. Instead of receiving donor eggs or undergoing a tricky ovary transplant, a woman could receive a transplant of stem cells and let nature take its course.

For now, research into the precursors of eggs has yielded more auspicious results, including the closest anyone has come to *creating* a virgin birth in mammals. In the 1980s, scientists had made the first attempts to create mice with two fathers or two mothers and, as we have seen, these experiments failed because of the way in which genes are imprinted – turned on or off by chemicals within the DNA. Trying to create a baby from two sets of DNA regardless of their origin went nowhere; instead of getting one 'dose' of a gene, the offspring usually ended up with double the amount from one parent and none from the other. Then, in 2004, came the breakthrough: Kaguya. Created by a group of Japanese scientists, Kaguya the mouse was named after a mythical princess whose true parentage was unknown – she was found inside a bamboo reed. In contrast, Kaguya the mouse's heritage could not have been better recorded. She was the first mammal to be born without a father and, what is more, the first animal in history to be born to two mothers.

The team, led by Tomohiro Kono at the Tokyo University of Agriculture, suspected that certain portions of the genome were posing the critical stumbling blocks when it came to imprinting. To circumvent these two problem regions, they realized they could turn to the biology of egg development. Remember that the genes silenced by imprinting are only silenced as the egg grows to maturity. By using DNA from an egg at an early stage of development, the scientists could gain access to these genes before they were locked. Kaguya was created from constructing an egg out of material from one mature egg and one immature egg, the equivalent of synthetic fertilization. Admittedly, Kaguya, like Dolly the Sheep before it, won a bit of a reproductive lottery. Out of the 371 reconstructed eggs that were implanted, only ten live embryos reached maturity, and only two survived outside the womb; Kaguya's sister was killed so that the genes involved could be studied in more detail. But within just three years, the scientists had honed the technology to produce fatherless mice that develop at a high success rate – equivalent, in fact, to the rate obtained with in vitro fertilization of normal embryos. Like Kaguya, these new generations of mice – all female, of course, since they only have sex chromosomes from eggs – have proved able to reproduce with males and produce fertile offspring.

To achieve this, Kono's team deleted two bits of DNA, called *H19* and *Dlk1-Dio3*, which are imprinted in the mother but also serve as key controllers of imprinting across the genome. The first imprinted gene to be identified, *IGF2*, is imprinted in the mother, and so is expressed in the child from the father's copy. What was particularly striking is that a substantial number of genes that have subsequently been discovered to be imprinted act as part of a pathway in which insulin-like growth factor-2 is crucial – the very thing that *IGF2* codes for. And one of these genes is *H19*.

The other critical imprinted region, *Dlk1-Dio3*, contains genes that encode proteins expressed only when they come from the father. The genes in the *Dlk1-Dio3* area are found throughout the embryo, but after birth, they are predominantly located in the brain, where their instructions for constructing the tiny pieces of machinery that regulate the workings of other genes do their work. These instructions are expressed only from the chromosome inherited from the mother. Some switch had to be turned, in order for the father's gene to stop influencing the offspring.

It was *H19* and *Dlk1-Dio3* that Kono and his team deleted to make Kaguya the mouse. Tampering with these sections effectively allowed them to use the egg's chromosomes as though they had come from a sperm.

◎◎◎

Making human babies using Kaguya-style genetic tinkering should be possible in the future. But doing so will yield only female offspring, unless we can get hold of a Y chromosome, even one manufactured in a lab. In 2007, a first step in this direction was taken: in a painstaking process, a synthetic chromosome was assembled using lab-made chemicals – that is, copies of the chemicals that make up DNA. The artificial chromosome contained 381 genes containing 582,970 base pairs – paired letters of the DNA alphabet. The pioneering biologist behind this construction was Dr Craig Venter, whose company, Celera Genomics, helped to unravel the sequence of the human genome, in parallel with the government-backed Human Genome Project, in 2003.

The initial design of Venter's artificial chromosome was based on a parasitic bacterium called *Mycoplasma genitalium*, which is considered the smallest naturally occurring genome in cell form. Venter's team extracted the bacterium's own DNA and inserted the synthetic reconstruction in its place. When they finally succeeded, they branded the creation as the first truly new artificial life form on earth. In Venter's words, the artificial chromosome was 'a very important philosophical step in the history of our species. We are going from reading our genetic code to the ability to write it.' Learning to write genetic code will be more complicated when it comes to creating artificial eggs and sperm, especially on the scale of *Homo sapiens*' twenty-three thousand genes, even after taking account of the 'non-coding' portions of the genome.

Still, the ability to create artificial eggs and sperm from stem cells is hailed as the technology that will finally bring an end to infertility. And rightly so, as it will also help us to uncover many of the remaining secrets surrounding how reproduction works. Experiments to make artificial eggs and sperm are likely to yield an increased understanding of genetic imprinting and the diseases that arise when imprinting goes awry. Since the cells that become the placenta can also be derived from these stem cells, this research could allow scientists to investigate how the early placenta develops and how disorders arise in it. And of course, artificial germ cells would allow individuals to bypass donors, avoiding the ethical issues of the egg and sperm trade. In fact, because the children produced from these cells will not be 'artificial' babies, scientists prefer to call them in vitro-derived cells.

In vitro-derived cells should be able to withstand freezing and be stored for future use, just as their 'natural' donated counterparts already are. The freezing procedures used today are much the same as they were in the early 1950s, when the modern technique was established. Scientists had been able to freeze and store sperm by the 1930s, but had not found a way to ensure that the sperm were not damaged, rendering them useless for reproduction. In 1949, two British scientists, C. Folge and A. D. Smith, were part of a team who finally succeeded in 'reviving' sperm after preservation, by using glycerol to maintain the sperm's structural integrity as it is plunged into temperatures around minus 196 degrees Celsius (minus 320 degrees F), and the process was improved substantially a few years later by the 'father of cryobiology', American zoologist Jerome K. Sherman. UK law currently only allows sperm frozen in this way to be kept for ten years, but as there is no evidence of any changes in quality over time, in theory, sperm suspended like this can last forever – no matter the source of the sperm.

Things are not as straightforward, however, when it comes to freezing eggs. Unlike sperm, which are quite small, the egg is big – the biggest cell in the body. When an egg is frozen, the large cell's greater fluid content often sustains ice-crystal damage. Even with an improved technique, known as vitrification, which flash-freezes the egg to avoid crystallization, there is only a ten percent success rate. If a young woman has her own eggs removed, chances are that when she is older and decides to use them, the process may very well have failed. And seeing that for women, removing eggs is quite invasive and uncomfortable, being able to make eggs in a lab from a woman's DNA is very attractive: it would mean that most of the process takes place outside of her body with far less risk.

So, how close are we? Bone marrow stem cells have proved extremely promising in the early experiments to create in vitro-derived germ cells: given the right signals, bone marrow stem cells are capable of becoming sperm. Further, three types of stem cells exist in bone marrow and it contains the cell-level blueprints for much of the body, including the heart, lungs, liver, kidney, bone, cartilage, fat, muscle, tendon, skin, and even the brain.

Regardless of the provenance of human embryonic stem cells, the proteins that act as signposts for developing male (sperm) and female (egg) cells can be detected in them. In fact, eggs have been made in the lab from both female and male embryonic stem cell lines. This is because the male embryonic stem cells have not yet expressed the *SRY* gene, which triggers the development of the testes and eventually the generation of sperm. Female embryonic stem cells, on the other hand, can only give rise to eggs. So without an artificial Y chromosome, women could only ever make artificial eggs, while men could make either artificial eggs or sperm. However, by keeping eggs fashioned from embryonic stem cells in a culture (that is, in a nutrient/chemical soup incubated with a prescribed mixture of gases at an appropriate temperature), scientists have been able to create parthenogenic embryos – embryos that begin to develop despite never having been fertilized by sperm. (Of course, this isn't all that peculiar when you consider that parthenogenesis is a relatively normal phenomenon; whenever eggs are kept in culture, they tend to start dividing on their own.)

In early 2006, two laboratories, one based in Germany and

the other in the UK, reported some remarkable results using embryonic stem cell lines. Earlier, scientists had successfully developed immature sperm, or spermatogonial stem cells (SSCs), from embryonic stem cells that, when they were injected into mouse eggs, developed into early embryos. The new research went one better. The teams transplanted SSC artificial sperm into the testes of mice that had no sperm of their own. After four months, the scientists observed sperm in some of these mice, generated from the transplanted cells. Unfortunately, the sperm did not move, or move very far, unaided; they weren't ever going make it to an egg. So to help the process along, the sperm were removed from the mice testes and injected into unfertilized eggs. Out of 210 eggs, 65 embryos were produced and transferred into surrogate mice mothers. Seven of these became baby mice, fathered by artificial sperm. But the baby mice sired this way were not very healthy, and they died at ages well below the average life expectancy of mice conceived naturally. Much remains to be worked out before artificial sperm are ready for humans.

Then in 2010, the first artificial human ovary was made, signalling a significant step towards the creation of eggs outside of a woman's body. The artificial ovary, built by researchers based at Brown University and the Women & Infants Hospital of Rhode Island, could move oocytes along the path to becoming mature eggs.

An ovary is a complex organ, composed of three major cell types, all of which need to be developed in a tissue structure for an artificial organ to function fully. According to Sandra Carson, the obstetrician and gynaecologist at Brown who led the team, the ovary provides not only a living laboratory for investigating how healthy ovaries work but a way to test for exposure to toxins and other chemicals that can disrupt egg maturation and health. In the future, Carson believes an artificial ovary might play a role in preserving the fertility of women facing cancer treatment – immature eggs could be salvaged and frozen before the onset of chemotherapy or radiation, and then matured outside the patient in an artificial ovary.

Carson's group has already used a lab-grown ovary to mature human eggs. To do so, they used a '3D Petri dish' made of mouldable gel, on to which two types of ovarian support cell could attach, forming a honeycomb structure. Seventy-two hours later, the third cell type was introduced, enveloping the immature egg cells in exactly the manner that would happen inside a real ovary. They managed to keep the structure healthy for up to a week. It is unclear whether the eggs developed in the artificial ovary contained all of the important genetic imprinting information, but Carson and her colleagues are performing further studies that they expect will make this possible. And even though it has not yet been dealt with in their investigations, the artificial ovary might equally well be used to mature artificial eggs too. In a university press release, Carson emphasized that the techniques are 'really very, very new' and that, setting aside her hopes for the future, it would be sensible to be cautious about where the experiments may lead.

◎◎◎

The obstacles to creating artificial sperm indistinguishable from healthy sperm created in a man are not so different, in some respects, to those faced in manufacturing mature eggs. Some of the labs that claim to have made sperm from bone marrow stem cells have made cells that can act as sperm but are not, strictly speaking, sperm. This makes a difference: the sperm 'actors' are basically little packages of DNA, and those sorts of packages can be extracted easily and directly from an infertile man's testes. There's little point in going through all the trouble of making sperm if they can't do more than that.

Sheffield University is home to one of the few labs around the world where artificial sperm is being manufactured properly. As part of this quest, Dr Allan Pacey, a male fertility specialist at the university, looks at semen on a day-to-day basis. In a very small room with a folding bed – installed for research purposes – Pacey explained the rigorous quality tests that are being used to understand what makes great sperm. In addition to the bed, Pacey's lab keeps a supply of porn magazines in its efforts to collect semen samples. Two thousand men from the Sheffield area have donated sperm to the study, which has verified that a count of twenty million sperm per one millilitre of semen is the norm for a fertile man – and that some men's semen contains no sperm whatsoever.

In the lab next door to the bedroom, a student scientist was busy with a microscope. Images of three or four sperm were projected from the microscope on to a computer screen. Seen this way, the sperm appeared huge and robust – around ten thousand times their true size. Normal sperm contain just one set of DNA, and all of these checked out as normal; none of them displayed the characteristic signs of carrying two sets: a grossly enlarged head or a head split in two. The day I visited, the lab was focused on something quite basic to reproductive function, and central to making artificial sperm that do more than serve as a container for DNA: they were measuring the length of normal sperms' tails.

The definition of a 'normal' tail length is not yet understood. Though the length of a sperm's tail affects how efficiently the sperm swims to the egg, one millilitre of a man's ejaculate will contain sperm displaying a huge variety of different tail lengths. So what length should an artificial sperm's tail be in order to have all the functionality of a real sperm? Making sperm, after all, is not just a case of making a cell with half as much DNA as the rest of our cells. 'What you are trying to replicate in the lab is not just a case of taking a cell and letting it divide. A skin cell could easily be made like that,' Pacey explains. To be a 'true' sperm, the cell has to move, and that requires a certain size and shape. In contrast, eggs, which start off round and remain that way, may be easier to construct. Pacey admits that, in theory, any cell might be altered to carry one set of DNA and be used to fertilize an egg in vitro, but that, he says, does not make a sperm or allow for natural conception using an artificial cell.

Another issue involves replicating how sperm develop in the body. Sperm do not develop in one spot but evolve progressively more mature forms as they move through the tubes of the testicles. It is in the final stages of sperm development that the head and tail are 'finished' and which pose such a challenge to researchers in the lab. The Sertoli cells, also known as nurse cells, guide the proper development and maturing of sperm in the testes, no matter where they are. In fact, scientists have managed to insert immature human and rat sperm into mice testes and observe it mature there, alongside the mouse's own sperm. Because human, mouse, and rat sperm look quite distinct, the researchers were able to confirm that nurse cells from one species can be used to mature sperm from another. In Pacey's view, Sertoli cells are an exciting avenue for investigation – they offer a way to use something like an 'artificial' testis to produce real human sperm.

These challenges are not impossible to overcome, and chances are that one or more of the three labs based in the UK and Japan that are competing to solve the problem will do it within a generation. Scientists expect that in vitro-derived germ cells will be ready to test in clinical studies within thirty years, and possibly sooner.

◎◉◎

Using in vitro-manufactured eggs and sperm could remove a significant issue that arises when a child is conceived from donor material: every bit of information that we can now glean from our genetics will be available to the parents to judge and assess and decide what to pass along to their children, including such things as whether or not they will be more susceptible to a disease than the general population. The egg and sperm banks that now offer their services often provide some information, particularly around superficial characteristics such as a donor's appearance and perceived intelligence: hair colour, eye colour, skin colour, and educational attainment. One solo parent specifically said that she had chosen a sperm bank because it released information about the donors' looks, character, and health. But often, people do not know their own family medical history – a mother or grandmother may have died long before the *BRCA1* gene could set her body on a path towards full-blown breast cancer. An artificial egg or sperm, in comparison, would come with a full genetic profile.

Using lab-made germ cells might also make us think more biologically about the family. When a person or a couple turns to a donor's egg or sperm, there's a niggling sense that biology has some lurking trump card to play. Will the child's genes harbour some undesirable disease or temperament? Will the child feel the tug of belonging, biologically, to another family, rather than the one into which he or she has been born? Are the children of a donor's eggs, or sperm, siblings in any sense? One British fertility doctor based in Oxford told me, for instance, that he thought the probability of two half-siblings meeting and marrying in the future was so slim that it was never something that worried him in his work. But this is the sort of taboo that still carries weight in society, just as the chances of it happening goes up with the number of IVF pregnancies. Dr Pacey of Sheffield University told me that he had been inundated with phone calls from infertile couples following a December 2010 press report in which it was claimed that the team at the North East England Stem Cell Institute had succeeded in creating a functioning artificial sperm. There is no doubt that the ability to make sperm and eggs in the lab has an allure for infertile individuals, for the very fact that it would sweep away the ambiguity inherent in the donor system – every bit of genetic identity will be their own.

If the option were to exist one day, the ultimate solo parent will probably be a woman who needs nothing but her own stem cells and an artificial Y chromosome to produce eggs and sperm. She might use two of her own eggs to create a child, converting one egg into a pseudo-sperm to fertilize herself, as scientists have already done in mice. And then, should an artificial womb become a reality, she might even forego pregnancy, allowing a doctor to set the ideal conditions for the foetus's development. She could even keep working, as men do, until the moment the baby is born.

This would be the great biological and social equalizer, a truly new way of thinking about sex. The question is not if it will happen, but when.

EPILOGUE

NEXT GENERATION

Over the past century, every innovation in reproductive technology, from the use of anaesthesia during childbirth to the first successful ovary transplant, has been met with criticism and resistance. Most have been seen as a threat to the traditional family – a change in the roles of men and women. But science is always conceiving the inconceivable – looking for the next frontier to cross.

In the 1950s, for instance, some scientists investigated the possibility of an all-female farm – a farm populated by only cows, sows, and ewes, with no bulls, boars, or rams. Animal breeding, as it has been conducted for millennia, requires raising and feeding big males for the sake of a little sperm. So why not create a line of virgin-born livestock that would improve farmers' profits and, as a bonus, yield new insights into genetics and sex itself? The assumption, and indeed the aim, of the work was that males are mostly expendable.

But this is not nearly the full story of reproduction. Going solo is not an option available to women alone. In fact, in a strange Aristotelian twist, the newest reproductive technologies – artificial wombs and artificial eggs – seem poised to give men more potential than women to make a baby without the opposite sex. And as shown in the studies of solo parents, the logistical route to parenthood – the *how* of having a child – is far less influential in a child's life than is the choice of parenthood – the *why*. Perhaps the necessity of our lifestyle and the ingenuity of medical science will force us to accept families that have been marginalized previously. They may even spur the emergence of families such as have never before existed, as genetics and biology are ripped out of the egg and sperm and allowed to be combined freely. One such case: two baby girls, called 'twiblings' by their mother, who were born in New York City in 2010. After she had gone through four failed rounds of IVF, the mother ended up using a donor's eggs, her husband's sperm, and two surrogate mothers (pregnant at the same time) to have her daughters, who are 'twins' only in the sense that they were born around the same time. Both surrogate mothers continued to be involved after the babies' births.

What were once invisible, indivisible seeds of life will act as portals rather than as ingredients to be brought together to create a child. The constituent, beautiful machinery of which eggs are built will very soon broker successful pregnancies and healthier babies – with three genetic parents. This fast advancing area of research involves transferring the male and female parents' DNA into a donor egg, which already contains a package of DNA from the donor, in a tiny organ in the egg called a mitochondrion. The child born from this process would inherit a fraction of his or her genetic code from the egg donor, breaking the rules of reproduction as we know them today.

There already are children born from permutations of

biological, gestational, and genetic input from more than two adults, and this will increasingly be humanity's future. Whether or not medical intervention soon gives women and men the choice of having biological children with a person of their same sex or completely on their own – unrestricted by the physical limits of the human body as well as the socio-economic limits on the human soul – the reproduction of the future is set to rewrite much of the fabric of human society. Male plus female equals baby will no longer be our only path forward. As we conceive the once inconceivable and take full control of how and when we bring the next generation into the world, we are sure to dislodge many notions of sex and gender along the way.

SELECTED BIBLIOGRAPHY

Prologue

Bondeson, Jan (1997). *A Cabinet of Medical Curiosities*. London: I. B. Tauris.

Porter, Ian Herbert (1963). 'Thomas Bartholin (1616–80) and Niels Steensen (1638–86) Master and Pupil'. *Medical History* 7(2): 99–125.

Wells, H. G. (1993). *Ann Veronica*, ed. Sylvia Hardy. London: Orion Publishing Group.

1. Planting the Seed

Andry de Bois-Regard, Nicolas (2010). *An Account of the Breeding of Worms in Human Bodies*. n.p.: Gale Ecco.

Aristotle (1910). *On the Generation of Animals*, trans. Arthur Platt. Oxford: Clarendon Press.

Bayrakdar, Mehmet (1983). 'Al-Jahiz and the Rise of Biological Evolution.' *Islamic Quarterly* Third Quarter: 307–15.

de Balzac, Honoré (1841). *Catherine de' Medici*, trans. Katharine Prescott Wormele (1894). Boston: Roberts Bros.

C., T. E. Jr. (1975). 'Galen on Why the Female Is More Imperfect than the Male.' *Pediatrics* 55(4): 562.

Gaziel, Ahuva (2012). 'Questions of Methodology in Aristotle's Zoology: A Medieval Perspective.' *Journal of the History of Biology* 45(2): 329–52.

Gordetsky, Jennifer, Ronald Rabinowitz, and Jeanne O'Brien (2009). 'The "Infertility" of Catherine de Medici and its Influence on

16th Century France.' *Canadian Journal of Urology* 16(2): 4584–8.

Gould, Stephen Jay (1991). 'Male Nipples and Clitoral Ripples' in *Bully for Brontosaurus: Further Reflections in Natural History*. New York: W.W. Norton, pp. 41–58.

Haimov-Kochman, Ronit, Yael Sciaky-Tamir, and Arye Hurwitz (2005). 'Reproduction Concepts and Practices in Ancient Egypt Mirrored by Modern Medicine.' *European Journal of Obstetrics & Gynecology and Reproductive Biology* 123(1): 3–8.

Hartmann, Franz (1891). *The Life and the Doctrines of Philippus Theophrastus, Bombast of Hohenheim, Known by the Name of Paracelsus*. New York: American Publishers Co.

Laqueur, Thomas (1994). *Making Sex: Body and Gender From the Greeks to Freud* (8th ed.). Massachusetts: Harvard University Press.

Tipton Jason A. (2006). 'Aristotle's Study of the Animal World: The Case of the kobios and phucis.' *Perspectives in Biology and Medicine* 49(3): 369–83.

Walker, William H. (2010). 'Non-classical Actions of Testosterone and Spermatogenesis.' *Philosophical Transactions of the Royal Society B* 365: 1557–69.

2. The Story of Safe Sex

Bittles, A. H. (2009). 'The Background and Outcomes of the First-cousin Marriage Controversy in Great Britain.' *International Journal of Epidemiology* 38(6): 1453–8.

Haldane, J. B. S. (1938). *Heredity and Politics*. London: George Allen & Unwin.

Hamilton, William D., and Marlene Zuk (1982). 'Heritable True Fitness and Bright Birds: A Role for Parasites?' *Science* 218(4570): 384–7.

Paul, Diane B., and Hamish G. Spencer (2009). '"It's Ok, We're Not Cousins by Blood": The Cousin Marriage Controversy in Historical Perspective.' *PLoS Biology* 6(12): e320.

Ridley, Matt (1995). *The Red Queen: Sex and the Evolution of Human Nature*. New York: Penguin Books.

Van Valen, Leigh (1973). 'A New Evolutionary Law.' *Evolutionary Theory* 1: 1–30.

Weissman, Charlotte (2010). 'The Origins of Species: The Debate between August Weismann and Moritz Wagner.' *Journal of the*

History of Biology 43(4): 727–66.

3. Desperately Seeking a Virgin Birth

n.a. (1955). 'Parthenogenesis in Mammals.' *The Lancet* 266–968.

Balfour-Lynn, Stanley (1956). 'Parthenogenesis in Human Beings.' *The Lancet* 267(6931): 1071–2.

Chapman, Demian D., Mahmood S. Shivji, et al. (2007). 'Virgin Birth in a Hammerhead Shark.' *Biology Letters* 3(4): 425–7.

Fournier, Denis, Arnaud Estoup, et al. (2005). 'Clonal Reproduction by Males and Females in the Little Fire Ant.' *Nature* 435(7046): 1230–4.

Hales, Dinah F., Alex C. C. Wilson, et al. (2002). 'Lack of Detectable Genetic Recombination on the X Chromosome during the Parthenogenetic Production of Female and Male Aphids.' *Genetics Research* 79(3): 203–9.

Iavazzo, Christos, Constantinos Trompoukis, Thalia Sardi, and Matthew E. Falagas (2008). 'Conception, Complicated Pregnancy, and Labour of Gods and Heroes in Greek Mythology.' *Reproductive BioMedicine Online* 17 (suppl. 1): 11–14.

Lutes, Aracely A., Diana P. Baumann, William B. Neaves, and Peter Baumann (2011). 'Laboratory Synthesis of an Independently Reproducing Vertebrate Species.' *Proceedings of the National Academy of Sciences* 108(24): 9910–15.

Olsen, M. W. (1965). 'Twelve Year Summary of Selection for Parthenogenesis in Beltsville Small White Turkeys.' *British Poultry Science* 6: 1–6.

Origen. Contra Celsum, Book 1. http://www.gnosis.org/library/orig_cc1.htm.

Spurway, Helen (1953). 'Spontaneous Parthenogenesis in a Fish.' *Nature* 171(4349): 437.

— (1955). 'Virgin Births.' *New Statesman and Nation* 50: 651.

Stouthamer, Richard, Robert F. Luck, and William D. Hamilton (1990). 'Antibiotics Cause Parthenogenetic Trichogramma (Hymenoptera/Trichogrammatidae) to Revert to Sex.' *Proceedings of the National Academy of Sciences* 87(7): 2424–7.

Strain, Lisa, Jon P. Warner, Thomas Johnston, and David T. Bonthron (1995). 'A Human Parthenogenetic Chimaera.' *Nature Genetics* 11: 164–9.

Watts, Phillip C., Kevin R. Buley, et al. (2006). 'Parthenogenesis in Komodo Dragons.' *Nature* 444(7122): 1021–2.

4. The Concert in the Egg

Hunter, Graeme K. (2000). *Vital Forces: The Discovery of the Molecular Basis of Life*. San Diego: Academic Press.

Lee, Yong Ho, Sung Gun Kim, Sung Hyuk Choi, In Sun Kim, and Sun Haeng Kim (2003). 'Ovarian Mature Cystic Teratoma Containing Homunculus: A Case Report.' *Journal of Korean Medical Science* 18: 905–7.

Masui, Yoshio, and Clement L. Markert (1971). 'Cytoplasmic Control of Nuclear Behavior during Meiotic Maturation of Frog Oocytes.' *Journal of Experimental Zoology* 177: 129–45.

Verkuyl, Douwe A. (1988). 'Oral Conception: Impregnation via the Proximal Gastrointestinal Tract in a Patient with an Aplastic Distal Vagina – Case Report.' *British Journal of Obstetric Gynaecology* 95(9): 933–4.

5. Secrets of the Womb

Brown, Jennifer R., Hong Ye, Roderick T. Bronson, Pieter Dikkes, and Michael E. Greenberg (1996). 'A Defect in Nurturing in Mice Lacking the Immediate Early Gene fosB.' *Cell* 86(2): 297–309.

Gupta, Anshu, Malathi Srinivasan, Supaporn Thamadilok, and Mulchand S. Patel (2009). 'Hypothalamic Alterations in Fetuses of High Fat Diet-Fed Obese Female Rats.' *Journal of Endocrinology* 200: 293–300.

Grant, Valerie J., and Lawrence W. Chamley (2010). 'Can Mammalian Mothers Influence the Sex of Their Offspring Peri-conceptually?' *Reproduction* 140(3): 425–33.

Isoda, Takeshi, Anthony M. Ford, et al. (2009). 'Immunologically Silent Cancer Clone Transmission from Mother to Offspring.' *Proceedings of the National Academy of Sciences* 106(42): 17882–5.

Jones, Helen N., Laura A. Woollett, et al. (2009). 'High-fat Diet before and during Pregnancy Causes Marked Up-regulation of Placental Nutrient Transport and Fetal Overgrowth in C57/BL6 Mice.' *FASEB Journal* 23(1): 271–8.

Reik, Wolf, and Jörn Walter (2001). 'Genomic Imprinting: Parental Influence on the Genome.' *Nature Reviews Genetics* 2: 21–32.

Villarreal, Luis P. (1997). 'On Viruses, Sex, and Motherhood.' *Journal of Virology* 71(20): 859–65.

Vrana, Paul B., Xiao-Juan Guan, Robert S. Ingram, and Shirley M. Tilghman (1998). 'Genomic Imprinting Is Disrupted in Interspecific Peromyscus Hybrids.' *Nature Genetics* 20: 362–5.

6. Out of the Test Tube

Deech, Ruth (2008). '30 Years: From IVF to Stem Cells.' *Nature* 454: 280–81.

Guttmacher, Alan F. (1943). 'The Role of Artificial Insemination in the Treatment of Human Sterility.' *Bulletin of the New York Academy of Medicine* 19(8): 573–91.

Heaton, Claude E. (1956) 'The Influence of J. Marion Sims on Gynecology.' *Bulletin of the New York Academy of Medicine* 32(9): 685–8.

Henig, Robin Marantz (2004). *Pandora's Baby: How the First Test Tube Babies Sparked the Reproductive Revolution*. Cold Spring, N.Y.: Cold Spring Harbor Laboratory Press.

7. Out of Time

Checa, Miguel A., Pablo Alonso-Coello, et al. (2009). 'IVF/ICSI with or without Preimplantation Genetic Screening for Aneuploidy in Couples without Genetic Disorders: A Systematic Review and Meta-analysis.' *Journal of Assisted Reproduction and Genetics* 26(5): 273–83.

Just, W., A. Baumstark, et al. (2007). 'Ellobius lutescens: Sex Determination and Sex Chromosome.' *Journal of Sexual Development* 1(4): 211–21.

Marais, Gabriel A. B., Paulo R. A. Campos, and Isabel Gordo (2010). 'Can Intra-Y Gene Conversion Oppose the Degeneration of the Human Y Chromosome? A Simulation Study.' *Genome Biology and Evolution* 2: 347–57.

Shah, Kavita, Gayahtri Sivapalan, Nicola Gibbons, Helen Tempest, and Darren K. Griffin (2003). 'The Genetic Basis of Infertility.' *Reproduction* 126(1): 13–25.

8. Real Men Bear Children

Alhusen, Jeanne L. (2008). 'A Literature Update on Maternal-fetal

Attachment.' *Journal of Obstetric, Gynecologic, & Neonatal Nursing* 37(3): 315–28.

Aristarkhova, Irina (2005). 'Ectogenesis and Mother as Machine.' *Body & Society* 11(3): 43–59.

Bulletti, Carlo, Antonio Palagiano, et al. (2011). 'The Artificial Womb.' *Annals of the New York Academy of Sciences* 1221: 124–8.

Otway, Nick, and Megan Ellis (2011). 'Construction and Test of an Artificial Uterus for Ex Situ Development of Shark Embryos.' *Zoological Biology* 30: 1–9.

Raju, Tonse N. K. (2011). 'Perinatal Profiles: Martin Couney and Newborn Infant Sideshows.' *NeoReviews* 12(3): e127–9.

Unno, Nobuya, Yoshinori Kuwabara, et al. (1993). 'Development of an Artificial Placenta: Survival of Isolated Goat Fetuses for Three Weeks with Umbilical Arteriovenous Extracorporeal Membrane Oxygenation.' *Artificial Organs* 17(12): 996–1003.

9. Going Solo

Kawahara, Manabu, and Tomohiro Kono (2009). 'Longevity in Mice without a Father.' *Human Reproduction* 25(2): 457–61.

Kenny, Nancy J., and Michelle L. McGowan (2010). 'Looking Back: Egg Donors' Retrospective Evaluations of Their Motivations, Expectations, and Experiences during Their First Donation Cycle.' *Fertility and Sterility* 93(2): 455–66.

Krueger, Andrew T., Larryn W. Peterson et al. (2011). 'Encoding Phenotype in Bacteria with an Alternative Genetic Set.' *Journal of the American Chemical Society* 133(45): 18447–51.

White, Yvonne A. R., Dori C. Woods, et al. (2012). 'Oocyte Formation by Mitotically Active Germ Cells Purified from Ovaries of Reproductive-age Women.' *Nature Medicine* 18(3): 413–21.

ACKNOWLEDGEMENTS

This book would not have been possible without the help and support of many people. I am indebted to you.

For inspiration, for your ears and eyes, for being intrepid archaeologists of obscure journals, for helping me pin down the science, and for your patience and many kindnesses: Professor Armand Leroi, Nalini Persad, Lynn Saliba, Dr Robin Lovell-Badge, Dr Evan Harris, Anjali Bhargava, Amanda Hargreaves, Frank Swain, Jessica Hamzelou, Dr Abul Tarafder, Dr David Mann, Mira Samlal and Surian Fletcher-Jones. For sharing your ideas on (some particularly difficult) ethics: Dr Anna Smajdor, Dr John Harris, and Clare Lewis-Jones MBE. For leisurely lunches in the once all-male common room at UCL, sharing J. B. S. Haldane stories and monographs, and contemplating the reproductive biology of the Blessed Virgin: Professor Sam Berry. For fascinating medical and scientific discussions: Professor Darren Griffin, Professor Karim Nayernia, Dr Olga Kapellou, and Dr Allan Pacey.

My very special thanks also to: Professor John Wood, whom I befriended on a flight to the Bahamas and who quite literally changed the course of my life; Peter Tallack, my literary agent; Marsha Fillion, who commissioned this book while at Oneworld, and my editor, Robin Dennis, who is pure genius; Tara Lumley-Savile, for putting up with me as I worked mornings, evenings, and weekends instead of doing fun things with you like the other mums – my deepest gratitude to you also because you sparked the idea that made me start this book; and Dr João Medeiros, who made me finish it. Thank you, thank you.

INDEX

ABOUT THE AUTHOR

Aarathi Prasad is a biologist and science writer. She has appeared on TV and radio programmes, including as presenter of Channel 4's controversial *Is It Better to Be Mixed Race?* and *Brave New World* with Stephen Hawking, as well as BBC Radio 4's *The Quest for Virgin Birth*, and written for *Wired*, the *Guardian*, and many other publications. Previously a cancer genetics researcher at Imperial College London, she subsequently moved out of the lab and into the worlds of science communication and policy, in areas including the passage in the UK Parliament of the Human Fertilisation and Embryology Act 2008. A single mother, Dr Prasad lives in London.